POPPER'S THEORY OF SCIENCE

Continuum Studies in Philosophy:
Series Editor: James Fieser, University of Tennessee at Martin

Andrew Fiala, *Tolerance and the Ethical Life*
Christopher M. Brown, *Aquinas and the Ship of Theseus*
Alexander W. Hall, *Thomas Aquinas and John Duns Scotus*
Robert Dyson, *St Augustine of Hippo*
Tammy Nyden-Bullock, *Spinoza's Radical Cartesian Mind*
Daniel Whiting, *The Philosophy of John McDowell*

POPPER'S THEORY OF SCIENCE

AN APOLOGIA

CARLOS E. GARCÍA

Continuum International Publishing Group
The Tower Building, 11 York Road, London SE1 7NX
80 Maiden Lane, Suite 704, New York, NY 10038

© Carlos E. García 2006

All rights reserved. No part of this publication may be reproduced or transmitted in any form or by any means, electronic or mechanical, including photocopying, recording, or any information storage or retrieval system, without prior permission in writing from the publishers.

Carlos E. García has asserted his right under the Copyright, Designs and Patents Act, 1988, to be identified as Author of this work

British Library Cataloguing-in-Publication Data
A catalogue record for this book is available from the British Library.

ISBN: 0-8264-9026-3 (hardback)

Library of Congress Cataloging-in-Publication Data

García, Carlos E.
    Popper's theory of science : an apologia / Carlos E. García.
      p. cm. – (Continuum studies in philosophy)
    Includes bibliographical references and index.
    ISBN 0-8264-9026-3 (hardback : alk. paper)
    1. Science–Methodology. 2. Probabilities. 3. Popper, Karl
Raimund, Sir, 1902– . I. Title. II. Series.
Q175.G314 2006
501–dc22
2005037553

Typeset by Aarontype Limited, Easton, Bristol
Printed and bound in Great Britain by Antony Rowe Ltd, Chippenham, Wilts

Contents

Acknowledgements	vii
Abbreviations	ix
1 Introduction	1
2 Solution to the Problem of Induction	8
3 Falsifiability	38
4 Corroboration	84
5 Verisimilitude	121
Appendix: List of Definitions	156
References	157
Index	162

Acknowledgements

My intellectual debts for this project are many and I may, inadvertently, fail to acknowledge some of them. To begin with, I wish to express my deepest gratitude to Chuang Liu for his continuous support and enthusiastic encouragement in almost all the stages of my work. I enjoyed our long discussions and benefited greatly from his advice and insight. I am also very grateful to Kirk Ludwig, with whom I had extensive discussions about several specific problems of this book. His valuable suggestions made my arguments more compelling and helped me to improve Chapter 5. Robert D'Amico has posed deep questions that led me to refine and better structure my views. My conversations with Robert A. Hatch about specific episodes in the history of science and their relationship with my topic were inspirational and prompted me to develop new ideas. Pablo R. Arango helped me to compile the index. None of them is responsible for the mistakes that still survive in this work

A significant part of this book was written while I was a doctoral student at the Department of Philosophy of the University of Florida, with the endorsement of the Fulbright Program. The Universidad de Caldas and the Universidad de Manizales freed me of teaching duties during the time I was engaged in research for this project. I also want to thank my family and my closest friends, especially Martin M. Steed and Luis E. García, for encouraging my interest in philosophy and for trying to make my life easier and less lonely.

Abbreviations

The main works of Popper are cited in the text as follows. (See References for publication details.)

CR	*Conjectures and Refutations: the Growth of Scientific Knowledge*
EPE	*The World of Parmenides: Essays on the Presocratic Enlightenment*
KBM	*Knowledge and the Body–Mind Problem: In Defence of Interaction*
LPS	*All Life is Problem Solving*
LScD	*The Logic of Scientific Discovery*
MF	*The Myth of the Framework: In Defence of Science and Rationality*
'Nv'	'A note on verisimilitude'
OK	*Objective Knowledge: An Evolutionary Approach*
OS	*The Open Society and its Enemies*
OU	*The Open Universe: An Argument for Indeterminism*
PH	*The Poverty of Historicism*
QSP	*Quantum Theory and the Schism in Physics*
RAS	*Realism and the Aim of Science*
'Rsr'	'The rationality of scientific revolutions'
SBW	*In Search of a Better World: Lectures and Essays from Thirty Years*
UQ	*Unended Quest: An Intellectual Autobiography*
WP	*A World of Propensities*

1
Introduction

This study deals with a particular portion of Popper's theory of science, namely its treatment of the problem of induction (including the proposal for a solution) and the notions of falsifiability, corroboration and verisimilitude. These four topics are the cornerstones of Popper's theory because, properly articulated, they are sufficient to account for the crucial problems of epistemology, not to mention that Popper's image of empirical science depends entirely upon them. In particular, they supply criteria to admit theories into the body of science or remove them from it; they provide means to select among competing theories and they give an explanation of the growth of scientific knowledge, including its piecemeal approach to the truth. In broader terms, they constitute an alternative to the ideas of logical positivism, offer support to scientific realism and give birth to an increasingly influential theory (now known as critical rationalism).

Given its cardinal importance for a methodology that admits hypotheses into the body of scientific knowledge with only the requirement of compliance with a formal criterion, special attention is given to the logical notion of falsifiability and to its relations with corroboration and verisimilitude. However, in spite of the simplicity that these four topics appear to have, this book does not attempt to analyse fully their many ramifications and complications. It settles for the less ambitious goal of providing a clear construal of each of them and explaining the basic form of their interrelation with the hope of giving the reader the tools necessary to grasp Popper's theory at its best.

Popper committed himself to a clear presentation of his ideas and encouraged others to scrutinize them severely with the implicit promise to abandon any view that did not resist criticism. This seems the appropriate attitude for someone who considered criticism to be 'the lifeblood of all rational thought'[1] and insisted on the fallible nature of the most sophisticated kinds of knowledge, science included. But the criticisms, which came from all quarters and appeared to be devastating, left the essence of his position untouched, and except for minor revisions and additional explanations here and there Popper did not change, let alone give up, any of his key notions. Since several authors[2] consider most of these criticisms successful (actually the myth of a conclusively refuted Popper is widespread), at some point one must ask why Popper did not modify drastically or renounce his central ideas after those criticisms were expressed. For many, this was an indication that despite his pronouncements to the contrary, Popper did not really admit criticism. I suspect that most of the criticisms were less apt than first thought. Although these alternatives hardly exhaust all the possible explanations, I am inclined to focus my discussion on them.

One must exercise caution, of course, when considering a popular view. In particular, one should not permit one's judgement to be affected by the degree of acceptance of the view under evaluation. For, as history shows with innumerable examples, the most fashionable views of the past have turned out to be shortsighted when they do not reveal themselves to be plainly false. This should be reason enough to avoid hastiness in embracing the most popular opinions of the last 30 years on the value of Popper's philosophy of science: namely that such a theory is a failed and hopeless project and that Popper's replies are simply desperate moves. Now, it is sometimes easier to join the stream of common opinion, and I must confess that I have frequently found pleasure rehearsing varied criticisms of Popper's philosophy. But as the tenets of critical rationalism advise us to expect, no conclusion can be considered so perfect that it should be exempted from further examination, and this must apply with unmitigated force not only to Popper's theory but also to the views of his many critics. In this book I endorse the opinion that most of the criticisms against Popper's theory of science are faulty and consequently that it is the right time to denounce their faults and vindicate such unfairly rejected theory.

I argue for that view by way of a twofold strategy. On the one hand, I attempt to disarm the most important objections, showing that they misconstrue Popper's intent. To this effect, I shall draw on material from the best-known critics (e.g. Kuhn, Lakatos, Feyerabend, Putnam and O'Hear) without neglecting some authors whose criticisms have received less attention in the literature. On the other hand, I shall propose interpretations that are more charitable to Popper's work and fit better in his overall philosophical project. I then must introduce some modifications to the concepts of corroboration and verisimilitude and argue for what I take is the right way to formulate the relations in which they stand one to the other while explaining what they contribute to a falsificationist theory of science. Nevertheless, all the revisions and modifications that I shall introduce are in the spirit of Popper's philosophy and are thus compatible with his whole theory. Let me state briefly the motivation for this book, explain my methodological approach and give a summary of the following chapters.

Motivation

A systematic defence of Popper's theory of science, by way of analysing its key components as identified above, constitutes a legitimate task for the following reasons:

- It fills a lacuna in the extensive critical literature on Popper's philosophy of science. Criticisms of the aforementioned notions are many and varied, but there are no systematic general defences, other than the rejoinders Popper himself wrote in several places in his work. So, it seems appropriate to engage in a project that gives a structured discussion of these notions, surveys the objections and rejoinders and provides support for a fair interpretation of Popper's theory.
- I believe that Popper's theory about the growth of science offers a good and persuasive account for today's science. However, his theory has been widely disputed, partly because of readings that interpret its key notions in isolation and take some of his views out of context. Therefore, a project that clarifies these concepts might contribute to dispelling misrepresentations that have pervaded and negatively affected the evaluation of Popper's theory of science.
- Popper introduced many of his controversial ideas in a rather sketchy way and suggested that they might be worth developing. I think a sympathetic reading of such ideas and a serious attempt to develop the notions that were presented too sketchily can contribute to a better understanding of Popper's project and perhaps promote additional contributions.

Method

There are two main approaches in this book: an expository approach and a constructive approach. In the expository approach, I discuss the appropriate background for each of the concepts that are essential to Popper's theory of science, so that once the concepts are introduced the reader is able to see how they fit in the broader structure. I consider different and alternative formulations of the key concepts as they appear in various places of Popper's work and stress the meaning of the changes whenever they emerge. Then I review the main objections to each of these concepts and I show how to interpret Popper's views so that it is possible to block or dissolve the criticisms argued for in the secondary literature. In addition, I shall explain why I disagree with the critics and try to locate my own position by reference to Popper's original notions. In the constructive approach, I reformulate the key notions (when needed) and attempt to give a coherent and workable characterization of Popper's leading ideas. If my reformulation is successful it will provide a solution for the most important problems identified by the criticisms as well as a more coherent and appealing picture of Popper's theory of science.

The material examined in this work has been organized in a very simple way. I have devoted one chapter to each of the four fundamental problems dealt with. Every chapter begins with a general discussion of the background required to understand its target notion. Then I proceed to the definitions and the changes introduced by Popper or demanded by flaws in the definition. After this, I give the important criticisms and my reply to them. There is some (logical) order in the problems and the notions discussed that might enhance our understanding of the interrelation among falsifiability, corroborability and verisimilitude. However, since I have emphasized the important points of connection repeatedly (both in the main text and in notes) the chapters can be read as self-standing units without significant loss of meaning. For clarification purposes and easy reference I have included also a list of the crucial definitions in an appendix. Those readers familiar with Popper's theory might want to skip the background discussion of each chapter and go directly to my presentation and assessment of the criticisms.

Main themes of the chapters

I devote Chapter 2 to a review of Hume's pioneering ideas on the problem of induction and examine Popper's solution to it. I survey Popper's construal of the problem of induction in different parts of his work, from the *Logic of Scientific Discovery* to volume I of the *Postscript*, stressing the changes that he introduces from time to time. Since he reads Hume as dismissing induction as a procedure that justifies reasoning from repeated instances of which we have experience to other instances of which we do not have experience (in everyday life as well as in empirical sciences) and emphasizes the need for a correct formulation of this problem, it is necessary to analyse carefully what he takes himself to be doing when he contends that he can *solve* the problem of induction.

As is known, Popper's solution to Hume's problem combines a dismissal of what he calls its psychologistic component with the adoption of falsifiability as the hallmark of empirical theories. Accordingly, I establish the connection between the theory of falsifiability and Popper's reply to Hume's challenge to the rationality of science. If Popper's solution is right, he is in the position of rescuing the very possibility and rationality of empirical science from the attacks of scepticism and relativism. On the other hand, since Popper commends a method that is entirely deductive and is capable of providing a good criterion of demarcation while securing that theories have informative content, we will evaluate his claims to that effect to determine whether his efforts to provide an account of empirical science without induction are successful. Once the topography of Popper's argument for that conclusion is laid down, I shall consider criticisms by some authors who do not consider Popper's solution to Hume's problem satisfactory nor his deductivism right.

Chapter 3 is concerned with the central topic of falsifiability. I show how this notion plays a key role in Popper's theory of science including the process of testing and theory selection. I explain the two definitions of falsifiability and the difference between falsifiability and falsification, essential for a correct understanding of many evaluative claims about theories and crucial to the dissolution of criticisms that point out that no falsificationist methodology of science can work. To do this, I explain the notions of basic statements, potential falsifiers, degrees of falsifiability, logical content and empirical (or informative) content, and I discuss the deductive testing of theories and the asymmetry between falsifiability and verifiability.

In addition, I shall develop Popper's suggestions on the role of falsifiability as a comparative notion together with his views on the epistemic nature of test-statements and their role in the feasibility of obtaining conclusive falsification. Since many of the alleged counter-instances to Popper's theory of falsifiability appeal to examples taken from the history of science, I shall show that the most overworked cases (e.g. the episodes of the scientific revolution) can be explained (and even interpreted better) from the point of view of Popper's philosophy. In section 3.3 of Chapter 3, I expound several objections to the theory of falsifiability by Kuhn, Lakatos, Feyerabend and Derksen, explaining how they might have originated. Since most of these objections are misconceived, I disarm them and show why they leave the theory of falsifiability unharmed.

In Chapter 4 I deal with the notion of corroboration, which gives us an idea of how well a theory stands up to severe tests, up to a certain moment. Since what Popper has written about corroboration is rather obscure and this notion is frequently confused with the notion of confirmation typical in verificationist epistemologies, I try to explain carefully why and how corroboration and verification are different concepts, and how to handle the difficulties that seem to arise whenever we attempt to apply Popper's notion of corroboration to the appraisal of scientific theories. In particular, I shall explain that although we can express the notion of corroboration by invoking concepts from the theory of probability, the notion itself is not probabilistic because it does not satisfy the axioms of the calculus of probability.

On the other hand, I introduce some distinctions between corroborability and corroboration emphasizing the contrast between modal and actual contexts. Furthermore, given the definitions of corroborability and falsifiability (in terms of testability) I explain how to establish a relation between both concepts in the right way so that we capture their logical connection and are able to use it to bolster Popper's theory of science. This demands a slight revision of the way Popper suggests the connection must be developed, but I think my amendment can correct some problems that Popper's suggestion generates. To put it plainly, Popper wants to establish a relation between falsifiability and corroboration that can dispel doubts about the so-called 'law of diminishing returns' (the view that after a while, the degree of falsifiability of a well-corroborated theory grows stale). However, Popper goes for a relation between falsifiability and corroborability that is easier to obtain since it is secured at the outset by the definitions. I shall correct this mistake and show that even with the relation between falsifiability and corroborability, Popper can have what he needs to keep his theory of science running. Once I do this, I examine criticisms of Popper's notion of corroboration by Putnam, Lakatos and Derksen. In particular, I devote special attention to the criticism (that I consider completely erroneous) that Popper sneaks induction back into his theory of science through his notion of corroboration. Finally, I use the previously introduced distinctions to neutralize such objections.

Chapter 5 is concerned with the theory of verisimilitude. Roughly speaking, this notion expresses the relative distance to the truth of a false hypothesis (or theory) and is important because it makes it possible to define progress for a sequence of false theories. In Popper's epistemology, the aim of science can be promoted by formulating theories that are closer to the truth than their overthrown competitors, and we recognize progress because our theories make successive approximations to the truth. Popper offered both a qualitative and a quantitative definition of 'verisimilitude'. For the sake of simplicity I am concerned mainly with the former.

The qualitative definition states that the verisimilitude of a (which stands for a statement but can be generalized to the case of theories) equals the measure of its truth content minus the measure of its falsity content, and that its degree of verisimilitude (say of a theory) increases with its truth content and decreases with its falsity content in such a way that if we could add to its truth content while subtracting from its falsity content we would be improving the theory. Several scholars quickly criticized this definition. In short, they claimed that it does not support ascriptions of verisimilitude of the kind that is useful to Popper's theory of science for cases in which we need to distinguish between two (unequally) false theories. Popper accepted this devastating criticism and suggested alternative ways to retain the spirit of the theory of verisimilitude without the technical components of the qualitative definition. I suggest a way to amend the qualitative definition and meet the challenge of the critics. My proposal builds on an overlooked part of Popper's notion of truth content. If my view is correct, I show that the qualitative definition works properly and secures the coherence of Popper's theory of science.

Finally, my book shows that Popper's method of conjectures and refutations, improved by means of critical and rational discussion, does not foster irrationalism in science or any other endeavour of the human spirit. On the contrary, while upholding induction might commit strict justificationists to an infinite regress of reasons, to scepticism, or to irrationalism, Popper's theory of science avoids these difficulties by construing science as a deductive enterprise and by rejecting foundationalism.

Notes

1. Popper 1974, p. 977.
2. Johansson 1975; Stove 1982; Burke 1983.

2
Solution to the Problem of Induction

Any moderately thorough account of Popper's theory of science must address his purported solution to the problem of induction. In fact, he has claimed as a main asset of his philosophy that it does away with induction while providing a rational account of empirical science. This, of course, is a very controversial claim for both friends and foes of the Popperian project. In this chapter, I examine the main lines of Popper's treatment of this important philosophical problem, to show how they relate to the rest of his theory. I also discuss some of the relevant criticisms of Popper's treatment of induction in the literature. I neither intend to give an exhaustive treatment of the problem of induction, nor to provide a detailed examination of the multiple objections that can be levelled at Popper's stance in this matter. Here, I aim at providing a suitable framework to my further defence of falsifiability as a desirable feature of scientific theories.

To prepare our way for this objective, let me begin by recasting Popper's treatment of induction and his understanding of the role that such a method plays in science. When the English translation of his *Logik der Forschung* appeared in 1959, Popper surprised the English-speaking philosophical community with the bold claim that he had produced a solution to the Problem of Induction.[1] That such a claim was destined to provoke heated debate can be illustrated by the fact that he had to explain over and over again what he regarded the problem of induction to be, and why he thought he had offered a solution for it. His initial treatment of this problem was rather simple and sketchy. He argued that current approaches to empirical science in the positivist tradition made induction the working mechanism of science. Sciences were adequately characterized by their reliance on something vaguely referred to as 'inductive logic'. But the known objections to induction raised by Hume sufficed to show that this characterization was on the wrong track. There was nothing that could be properly called 'inductive logic' and no way to ground the truth of general statements on the truth of particular ones. Hence, Hume's objections put epistemologists in the very uncomfortable position of being unable to account for the rationality of scientific knowledge if they were to maintain that empirical sciences were based on induction. According to Popper, the problem can be overcome if, instead of searching for well-established inductive support (a hopeless endeavour, in his opinion) one realizes that for demarcation and truth-decision purposes, falsification gives what verification promised but could not deliver.

The formulation of a criterion of demarcation between empirical and non-empirical theoretical systems received a novel treatment. Let me illustrate this by mentioning three points that are amenable to discussions current in the philosophy of science of the 1930s. In the first place, Popper explained that the criterion of demarcation between science and metaphysics should not be construed as a criterion of meaning, since we certainly can find meaningful theories on both sides of the dividing line. In the second place, he charged that the principle of verifiability could not provide a suitable criterion of demarcation, as the unwelcome consequences of its application in the days of the logical positivists blatantly showed. In the third place, he thought that a simple suggestion was powerful enough to dissolve all the difficulties of the positivist accounts of science: the undisputed fact that no general statement is verifiable but that they all are falsifiable. On Popper's view, falsifiability was an appropriate criterion of demarcation and became the hallmark of empirical theories. The notion proved to be so fertile that it originated a whole theory of science. But how does the solution to the problem of induction come into play here? The traditional (positivist) treatment of science made induction the distinguishing characteristic of empirical theories, but was unable to reconcile the evident fact of the progress of science with the challenges to rationality raised by Hume's criticisms of induction. Falsifiability came to the rescue by furnishing a deductive answer to the question of demarcation, and by providing an explanation for the progress of science in non-justificationist terms.

I do not have to mention that this story was not convincing enough for most philosophically minded readers. Many were unable to see a solution to the problem of induction in Popper's proposal. Others expressed various complaints about the tenability of Popper's solution. In their view, the problem of induction was still alive and well.[2] In addition, they pointed out what they considered were fatal flaws in Popper's theory of science. By way of example, let me mention two general criticisms of Popper's solution of the problem of induction. Howson (1984) contends that Popper failed to develop a tenable solution to this infamous problem,[3] if the assumption that all universal statements have a zero degree of probability is correct. In Howson's view, the way Popper approaches the question prevents him from meeting the requirement of epistemological neutrality that he wishes to impose on others, and makes the choice of theories or hypotheses non-rational. A second criticism, in some way licensed by Popper's own remark about a 'whiff of inductivism', says he must give some room to induction in his theory of science if he wants to account for the successes of scientific prediction. According to these critics, Popper reinstates induction into his theory of science through the back door. They mean that only by assuming that a theory well corroborated will perform equally well in future tests (minimally in the very same tests it has successfully withstood so far) can we make sense of science and understand its predictive features. But this is just a way of saying that past performances are linked to future performances, and thus relies on an inductive type of argument. We might be in a better position to appraise the force of both kinds of criticism, if we follow closely the different formulations of the problem of induction advanced by Popper. But before doing that, let us recapitulate briefly Hume's original treatment of this problem.

2.1 The problem of induction

As is widely known, the problem of induction was discovered by Hume, who was unhappy with our predisposition to believe that: (1) the truth of universal statements could be justified by the truth of particular statements; (2) the future occurrence of events resembles the past occurrences of the same (type of) events; and (3) it is possible to establish a cause–effect relationship between events which occurrence has been invariably associated in the past. The last statement is particularly important, since Hume himself did not use the word 'induction' in any of his arguments. Rather, he was concerned, in a broad sense, with questions that fall under one or other of the above headings, and spent a good deal of time examining the cause-and-effect relationship. According to Hume, the objects of human reasoning fall into two types: 'relations of ideas' such as the truths of mathematics, which are either 'intuitively or demonstratively certain', and 'matters of fact', such as the claims of science and common sense, which are highly likely or extremely probable, at best. The indubitable truth of relations of ideas can be discerned by analysing the conceptual interrelationships among its constituent terms. By contrast, the truth of matters of fact can only be established by means of concrete experience and involves nondemonstrative arguments which outcome can never be rendered absolutely certain. Roughly speaking, the claim that the sun will rise tomorrow, based on our inductive generalization from past experiences, is not better than the claim that it will not rise; for neither of them implies contradiction.[4] But we cannot demonstrate the falsehood of the second claim as we can reasonably demonstrate the falsehood of $3 \times 5 = 30$.

Inductive inferences from experience do, indeed, presuppose that the future will resemble the past in key relevant aspects. However, Hume's objections show the theoretical possibility 'that the course of nature may change, and that the past may be no rule for the future', which would render prior experience inferentially useless.[5] Although knowledge claims based on what we have experienced in the past are usually acceptable, the problem arises when we try to justify conclusions about the present or the future, inferred from the findings of past experience. No matter how certain those findings may be, we are not in the position of justifying the move from past experience to future results in order to substantiate a claim of knowledge of the kind: 'The occurrence of phenomenon x has always been followed by the occurrence of phenomenon y.' Something similar can be said concerning predictions involving future experience.[6]

Despite the difficulties of justification, however, the common mind resorts frequently to arguments that take for granted the relation between past and future. Since these inferences are not completely arbitrary, one may suspect that they are determined by some principle. Hume identified it as *custom* or *habit*: 'For wherever the repetition of any particular act or operation produces a propensity to renew the same act or operation, without being impelled by any reasoning or process of the understanding, we always say, that this propensity is the effect of *Custom*.'[7] It is this principle that impels us to make inferences from 'a thousand instances' when we do not dare to do it from a single instance, and that leads us to expect the appearance of an object (i.e. heat after flame) from the appearance of another that has been constantly united with it in the past.

On Hume's view, our mind uses three principles to establish a connection among the different thoughts that are presented before the memory and the imagination: resemblance, contiguity and cause–effect. The first two of these are intuited through sensation or reflection; the third one is more complex. Hume offers two (very similar) definitions of the term 'cause'. According to the first, any pair of objects such that one is followed by another where all the objects similar to the first are followed by objects similar to the second, instantiate the cause–effect relation. He states the second definition appealing to the appearance of one object that always conveys the thought of another.[8] Hume's analysis of causation, whose connection with the problem of induction arises from his view that all reasoning concerning matters of fact is founded in the relationship between cause and effect,[9] goes far beyond the notions of spatial contiguity, temporal succession and joint occurrence. He notes that we have no more than probable knowledge of the relationship between causes and effects. He suggests that, on the same count, reasonings regarding matters of fact, being founded on a species of analogy that must be inductive and probabilistic, never yield absolutely certain conclusions.

Hume also considered the belief in the cause–effect relationship as a natural outcome of a species of natural instinct that no reasoning can avoid: a habit of causal expectation so universally extended that without it men could hardly have survived. This kind of expectation, of course, reveals our faith in the regular concurrence of two events, constantly reinforced by the high frequency of their occurrence. Unfortunately, it does not justify our cause–effect inferences. In fact, we cannot appeal to the high probability of an occurrence, since this feature is itself based on induction, and cannot be invoked as a reason to believe in causation.[10]

To sum up, Hume has shown that a sceptical conclusion is inevitable. On the one hand we cannot escape from the burden of causal expectation and on the other we cannot see that such a habit provides any reason whatsoever to justify the belief in causal connection. If Hume is right about the origin of the idea of necessity, and we accept his remark according to which 'all probable reasoning is nothing but a species of sensation',[11] it looks as if we are committed to a totally sceptical answer to the general problem of justifying induction. Let me emphasize that Hume did not deny that there are regularities in the world; he simply showed that there were no logical links between different events occurring at different times. He also questioned the attribution of causal relationships to successive events, since the fact that x happens before y does not necessarily means that x causes y. Hume's sceptical conclusions do, in fact, challenge every kind of nondemonstrative argument, whether or not grounded in causal imputation. Since any attempt at justifying logically the inference from particulars to universals is deemed to fail, we are left in a bad position to give a good account about our knowledge of general facts.[12]

2.2 Popper's solution

As I have just illustrated, under the classical treatment of induction one finds a handful of distinct although related philosophical problems. One of these can be recast as 'Why, if at all, is it reasonable to accept the conclusions of certain inductive arguments as true – or at least as probably true? Why, if at all, is it reasonable to employ certain rules of inductive inference?' Popper is concerned mainly with this problem, and thinks that the question: 'are inductive inferences justified?' encapsulates its important epistemological aspects.

Since there is nothing like 'the' problem of induction, but many problems grouped under this heading – though Popper offers a response to all of them – there is no single approach and, presumably, no single solution. In what follows the reader can see that what Popper does, in effect, is to give an account of science which can coexist with a position that denies there can be evidential support or justification for law-like claims, that also denies there can be projection of properties, and finally that denies we can judge scientific theories by their 'track record' or by appeal to empirical evidence directly – as in direct acquaintance or sense-data accounts.

In *LScD* Popper entertained three alternative questions that characterize the problem of induction. (1) Are inductive inferences justified? (2) Under what conditions are inductive inferences justified? (3) How can we establish the truth of universal statements which are based on experience? Popper agrees, in general, with the core of Hume's objections, but he dislikes Hume's psychologistic explanation of induction and causality. What Popper finds right in Hume's account pertains to noncontroversial matters. For example, let us consider a purported principle of induction: that is, 'a statement with the help of which we could put inductive inferences into a logically acceptable form'.[13] Such a principle, Popper says, cannot be analytic (if it were so, inductive inferences would be just logical transformations exactly like their deductive counterparts); hence, it must be synthetic. But a principle like this requires a justification of its own, and putting aside the unlikely scenario in which one accepts an *a priori* justification (an avenue difficult to reconcile with the basic premises of empiricism) the justifying principle would have to be inductively justified at its turn. Soon, we would be trapped in a *regress*:

> For the principle of induction must be a universal statement in its turn. Thus, if we try to regard its truth as known from experience, then the very same problems which occasioned its introduction will arise over and over again. To justify it, we should have to employ inductive inferences; and to justify these we should have to assume an inductive principle of a higher order; and so on.[14]

Popper concludes that the various difficulties of inductive logic are insurmountable and cannot be alleviated by claiming that inductive inferences '*can attain some degree of "reliability" or of "probability"*', since taking the principle of induction as probable rather than true generates similar problems. As mentioned above, the examination of the problem of induction in terms of logical justification reveals two undesirable outcomes: the impossibility of justifying law-like statements which transcend experience (something we would need to do, if we accept induction); and the untenability of a verificationist principle of empiricism, according to which only experience enables us to establish the truth of universal statements. Popper's solution of the problem of induction goes in two steps: the first consists in eliminating psychologism by withdrawing attention from the process 'of inventing or conceiving a theory', because it is not amenable to logical analysis; the second consists in formulating an alternative method to test empirical theories. Let us see briefly how the first strategy can be implemented.

According to Popper, good epistemology should analyse the logical justification of statements and theories instead of concerning itself with the facts that surround instances of discovery. This contrast is customarily discussed in terms of the difference between the context of discovery and the context of justification. The former supplies the details in the process of generating a new theory and is sometimes used to explain novel breakthroughs in many fields (literary creations, musical or scientific achievements) recurring to the psychological phenomena that precede them. History of science has plenty of anecdotes (some of them fantastic) that tell us how a particular thinker came to conceive a hypothesis or an ingenuous solution to a problem. But, however inspiring or colourful these anecdotes may be, they are extraneous to Popper's theory of science. In this theory one has to concentrate solely in the context of justification, and try to disentangle the logical relations among the diverse components of a theory. Furthermore, when tackling this objective, one does not carry out a 'rational reconstruction' of the steps taken by the scientist to get a discovery or advance a new truth, because the processes related to the forming of 'inspirations' belong to empirical psychology rather than to the logic of knowledge. By contrast, one can carry out a reconstruction of the tests that prove that an idea yields knowledge or an inspiration constitutes a discovery. In this sense, Popper argues, there is neither a logical method to obtain new ideas, nor is it possible to reconstruct such a process with the help of the logic of scientific discovery, and the psychological process of invention is not very important. In fact, there is an irrational element in all discoveries ('a creative intuition') and its analysis and characterization are beyond the object and scope that Popper sets for his epistemology.

Popper's remarks about the leading role of the context of justification – the Kantian *quid juris* – for epistemology prepare the road for the suppression of induction and its replacement by the deductive way of testing, which constitutes the second step mentioned above. The new method of theory-testing is applied in four distinct and yet complementary ways. In the first one, the conclusions of the theory, hypothesis or anticipation that one intends to test are deduced, and the conclusions are compared among themselves, with the aim of determining the *internal coherence* of the system. In the second way, one determines the *logical form* of the theory to decide whether or not its study is the business of empirical science, since many theories (e.g. tautological ones) are not the object of the *logic of discovery*. In the third way, the theory to be tested is compared with other theories to determine if it constitutes a *scientific advancement* once it has withstood successfully attempts to refute it. In the last way, we need to be concerned with checking the empirical applicability of the conclusions derived from the theory in order to find out if the new consequences of the theory stand up to the demands of practice. As in the three previous ways, one only needs to appeal to a deductive procedure of testing. This way of testing can be sketched as follows:

> With the help of other statements, previously accepted, certain singular statements – which we may call 'predictions' – are deduced from the theory; especially predictions that are easily testable or applicable. From

among these statements, those are selected which are not derivable from the current theory, and more especially those which the current theory contradicts. Next we seek a decision as regards these (and other) derived statements by comparing them with the results of practical applications and experiments. If this decision is positive, that is, if the singular conclusions turn out to be acceptable, or *verified*, then the theory has, for the time being, passed its test: we have found no reason to discard it. But if the decision is negative, or in other words, if the conclusions have been *falsified*, then their falsification also falsifies the theory from which they were logically deduced.[15]

We have here a new image of empirical science: a view that does not appeal to inductive principles to establish the desired connection between theories and experience. Having rejected induction, turning to deduction seems the natural move. Popper claims that the use of deductive methods of testing in addition to some regulative principles (e.g. falsifiability and corroboration) suffice to explain the dynamics of empirical science, provide a demarcation criterion (something that the inductivists failed to do) and give an answer to the fundamental question of epistemology; namely, how does the process of the growth of knowledge take place? According to Popper, to give an exhaustive treatment of these questions, eliminating at the same time the difficulties posited by an inductivist logic, we only need the deductivist machinery just outlined. Let us turn now to his characterization of the logical problem of induction in *Conjectures and Refutations*.

In other words, the logical problem of induction arises from (*a*) Hume's discovery that it is impossible to justify a law by observation or experiment, since it 'transcends experience'; (*b*) the fact that science proposes and uses laws 'everywhere and all the time' ... To this we have to add (*c*) the *principle of empiricism* which asserts that in science, only observation and experiment may decide upon the *acceptance or rejection* of scientific statements, including laws and theories.[16]

Popper thinks that Hume was right when claiming that there exists no relation of a logical character between observational evidence and theories or hypotheses on which their truth could be established or even made likely. This leads to the extreme form of scepticism we are familiar with. However, Popper has a recipe to avoid this scepticism: we only need to appreciate the significance of the fact that observational evidence may falsify a law-like statement that asserts the existence of regularities. Given that our aim in science (as in many of our daily activities) is truth, we should prefer a theory that has not been falsified and thus which, for all we know, may be true, over one that has been falsified by observation. On these lines, scientific practice consists largely of vigorous and continuous attempts to falsify the theories that have been proposed to solve the different problems in which the scientist is interested. Theories that, at any given moment, have so far survived such attempts are the ones that we should accept (on a tentative basis and for further experimental testing) and provide us with the best solutions to these problems. Yet, the status of all our scientific knowledge is that of a conjecture, a tentative solution to a problem that is accepted while it stands up to the tests. This new visit to the problem of induction is enriched by Popper's emphasis in the conjectural character of all our scientific knowledge. This was not, however, his last word on induction. He came back to this problem in his influential *Objective Knowledge*.

In this book, Popper contends that what makes it possible to find a solution to the traditional problem of induction was his reformulation of it. Moreover, he distinguishes between a logical and a psychological formulation of the problem of induction in Hume's writings. This distinction is important, since Popper now presents his solution as a dismissal of Hume's psychological formulation and as an argument for a different answer to the question of the truth of universal theories (which he takes as part of the logical formulation). The two problems in Hume's original work can be put as follow: (1) the logical problem of induction, namely the question of whether we are justified in inferring conclusions about things of which we have no experience based on instances of which we have experience; and (2) the psychological problem of induction, namely that of explaining why reasonable people expect and believe that 'instances of which they have no experience will conform to those of which they have experience'.[17] Popper agrees with Hume's negative answer to the logical problem of induction but disagrees with his conclusions about knowledge and argument, and with the way he developed the consequences of his psychological formulation of the problem. In particular, Popper is dissatisfied with the thoroughgoing irrationalism of Hume's psychological explanation of induction. Furthermore, Popper thinks that he can overcome most of Hume's difficulties by avoiding any talk in subjective terms (beliefs, impressions and justifications of beliefs) and referring instead to statements, explanatory theories and observation statements, and asking whether the claim that a theory is true can be justified. The logical problem of induction (that Popper labels L), restated in an objective or logical mode of speech, looks quite different now:

(L) Can the claim that an explanatory universal theory is true be justified by 'empirical reasons'; that is, by assuming the truth of certain test-statements or observation statements (which, it may be said, are 'based on experience')?[18]

Popper's answer is the same as Hume's: 'no'. We cannot justify the claim that an explanatory universal theory is true based on the truth of a set of test-statements, no matter how big its cardinality may be. As we can see, there is no major difference between Popper and Hume on this score, other than Popper's insistence on the disastrous consequences for empirical science if Hume's challenge remained unanswered: there is no way out of irrationalism in so far as Hume's arguments prevail. Fortunately, Popper thinks, his objective formulation of *L* enables us to bypass irrationalism by simply taking into account that the assumption of the truth of test-statements sometimes allows us to justify the claim that an explanatory universal theory is false. In other words, since universal statements are falsifiable, whereas particular statements are verifiable, we can exploit this asymmetry to explain how universal theories are decidable with the help of particular (verifiable) statements. The result is that scientific theories cannot be established as true but can be refuted, and if they are so, they may foster the progress of knowledge by telling us where to look for promising lines of further work. In this process the required appeal to experience secures the link between theories and states of affairs, and the hopeless problem of deciding the truth of universal statements can be fruitfully replaced by the problem of selecting between competing theories and choosing the theory whose falsity has not yet been established. On the other hand, Popper introduces a new element that was absent in the old days of *LScD*, namely the notion of verisimilitude. This notion furnishes him a practical way to block the negative result of Humean scepticism. One no longer has to worry about demonstrating that universal theories are true. Under the principles of Popperian epistemology it suffices to show that they may be truth-like: that is, that they may get closer to the truth.[19]

Compared to the very condensed arguments against the possibility of inductive logic which are advanced in *LScD*, Popper's careful distinction between the logical and the psychological formulations of the problem of induction in *OK* is more illuminating. Furthermore, he explains clearly what he takes himself to have done to solve the problem of induction, and what that solution amounts to. Both points can be summarized by means of the consequences of Popper's positive proposal (i.e. use deduction to show that some explanatory universal theories are false) which can be further clarified if we place the circumvented or solved difficulties under four headings: (1) the distinction between rational and irrational statements: Popper's deductive theory of testing allows us to distinguish between some of the ravings of the lunatic – in so far as they are refuted by test-statements – and the theories of science, which are best fitted to pass demanding tests; (2) the acceptance of the *principle of empiricism*: we can decide about the truth or falsity of factual statements merely by resorting to experience, although only judgements of falsity are conclusive; (3) the agreement with the methodology of empirical science:

deductive testing of theories is perfectly compatible with science; (4) the identification and further dismissal of the psychological component of the problem of induction: according to Popper's view, Hume's answer was wrong. A correct formulation of the traditional problem of induction exhibits the weaknesses of its psychological formulation and shows that psychology should be a part of biology instead of a part of epistemology.[20]

An even more extended treatment of the problem of induction appears in *Realism and the Aim of Science*, the postscript to *LScD*. The whole Chapter 1 of the book is devoted to a lengthy and detailed examination (which runs for about 150 pages) of the problem of induction and related topics. I will not report on the minutiae of Popper's discussion of these topics. I shall limit myself to survey briefly the main arguments advanced there. Popper distinguishes four phases of the discussion of the problem of induction, the first three of a logical or epistemological or methodological character, the last one of a metaphysical character. The first phase is a slight variant of a formulation presented earlier. It is called Russell's challenge and can be phrased in the following terms: *what is the difference between the lunatic and the scientist?*[21] According to Popper, this phase is closely related to the 'problem of demarcation', or to the problem of finding an adequate criterion to distinguish scientific (empirical) theories from other kinds of theories, although it can also give a characterization of what a 'good' theory consists of by contrast with a 'bad' theory. The second phase addresses the so-called '*problem of rational belief*', which is the problem of finding a way to justify (rationally) our belief in theories that have been tested by observations and have escaped falsification so far. On Popper's own words, this phase is 'less fundamental and interesting' than the previous one and it arises from an excessive – perhaps unwarranted – interest in the philosophy of belief. The third phase relates to the question of whether we may draw inferences about the future or whether the future will be like the past: something Popper calls '*Hume's problem of tomorrow*'. Being no more than a 'philosophical muddle', it is even less fundamental than the second phase. The last phase concerns the metaphysical problem of 'tomorrow', or '*the fourth or metaphysical phase of the problem of induction*'. It is different from the previous phase because it is usually formulated by means of existential statements that refer to the existence of laws of nature or the truth value of universal statements regarding invariant regularities.

Popper maintains that the central problem of the 'philosophy of knowledge' is that of adjudicating between competing theories. This, which is an urgent practical problem of method, leads to the question of justifying our theories or beliefs, that, in its turn, requires a characterization of what a 'justification' should consist of, as well as a decision concerning the possibility of offering a rational justification of theories or beliefs. If providing a rational justification means offering positive reasons (e.g. appeal to observation), Popper denies that there is such a possibility, since he is not a justificationist. On Popper's view, his theory of science yields a satisfactory solution to the first problem, and he thinks that the assumptions which lead from the first problem to the second are incorrect because we can not offer any positive reasons for holding our theories as *true*. Accordingly, since the belief that we can give reasons to justify an attribution of truth to a theory is neither rational nor true, we can reject the second problem – the justification of theories – as irrelevant.

Some readers have contended that this refusal to give positive reasons for holding a theory true commits Popper to scepticism about science and to a thoroughgoing irrationalism, but the charges are unwarranted. For Popper, although there may be shared views between the sceptic about science and himself (they both deny that it is possible to produce positive reasons for holding a theory true), there is a big difference which cannot be overlooked. Unlike the irrationalist, he gives an affirmative solution to the problem of deciding when and why one theory is *preferable* to another, where by 'preferable' he means being a closer *approximation to the truth*. Popper holds that we can have reasons to think or conjecture, for a particular theory, that this is the case. And this is precisely his solution to the first phase of the problem of induction.[22] One that replaces the old challenge posited by the problem of justification by a totally different problem, namely that of explaining or giving critical reasons to support our preference of a theory to one or various competitors, or 'the problem of *critically discussing* hypothesis in order to find out which of them is – comparatively – the one to be preferred'.[23]

I shall defer my analysis of the notion of 'being closer to the truth' for Chapter 5 below. For the time being, let me just point out that Popper makes a great deal of this notion for his theory of science. It is particularly telling that he discusses it in the light of what he calls 'rational criticism', an attitude that should prevail in good philosophy and epistemology, and that replaces the problem of justification by the problem of criticism. The point is that any decision regarding the epistemological status of a theory, and especially its acceptance or selection over its genuine competitors, has to be supported by critical examination of its ability to solve the problems it is designed to solve, and the way it stands up to the tests. Most importantly, when critically examining a theory we do not attempt to produce valid reasons to prove that the theory is true (which was the first phase of the problem of induction), but we attempt to produce valid reasons against its being true (we attempt to overthrow the theory), or in some cases (e.g. when we have not been able to overthrow the theory) against the truth of its competitors. In all these cases truth, which plays the role of a regulative idea, is one of the standards of criticism.

Another important feature of rational criticism is that, unlike justificationism, it makes no claim whatsoever pertaining to the final character of reasons to establish the falsity or truthlikeness of a theory under examination. For

Popper, rational criticism is the means by which we 'learn, grow in knowledge and transcend ourselves', that is, it is the most productive tool of his theory of science, but it does not furnish final, let alone demonstrable, reasons. On the contrary, according to Popper's epistemology there are no ultimate reasons: all reasons are themselves open to criticism, all reasons are conjectural, hence they can be examined infinitely, and this alone distinguishes them from any kind of justification. By contrast, those who defend the existence of justificatory statements sooner or later find themselves committed to ultimate reasons, that is reasons which cannot be justified themselves (unless they accept the idea of an infinite regress).

Adoption or rejection of theories is governed by the results of critical reasoning in combination with the results of observation and experiment as is demanded by the principle of empiricism. In the course of searching for increasingly better theories we start by advancing tentative conjectures, in the hope that after testing they might turn out to be true. To assure that we select only the best theories available, we devise increasingly severe tests which help us to decide on the fitness of those theories that survive testing. On the other hand, Popper also has an answer for those who think that induction is required to account for our ability to learn from experience and even to obtain 'experience'. In fact, they seem to think with Hume that exposure to the repetition of events affords inductive reasons that persuade people of the existence of regularities in the world, and that inductive inference is the only way to learn from experience. Popper denies neither the existence of regularities in nature, nor is he opposed to the view that one can learn from experience; he just thinks we do not discover the regularities by any type of inductive procedures but by 'the essentially different method of trial and error, of conjecture and refutation, or of learning from our mistakes; a method which makes the discovery of regularities much more interesting than Hume thought'.[24]

And it is precisely this method (trial and error) that generates 'experience'. One does not learn from inductive inferences; one learns from one's errors, by critically examining observations and the results of experiments, thus learning where one has been mistaken. So the role of observation in the acquisition of learning is not that of exhibiting regularities by repetition but that of falsifying claims, making clear by this way that we can draw inferences from observation to theory, inferences which establish that a particular theory or a hypothesis is false while another may be corroborated by observation.[25]

In regard to the second phase of the problem of induction, that of showing the reasonableness of our belief in the results of science, Popper contends that it is less fundamental than the first phase, since the theoretician who is interested in the appraisal of theories can do his job without entertaining any belief about them, for he only needs to consider a good theory as one which is important for

Solution to the Problem of Induction 21

further progress. On the other hand, Popper takes the philosophy of belief as mistaken in light of his notion of objective knowledge. The real problems of epistemology, he contends, are not those of explaining *subjective* beliefs but those related to the *objective* contents of the beliefs. If we take the epistemological import of this distinction at face value, then we have to concede that the study of scientific theories has no bearing on the study of beliefs. Moreover, since a theory may be true although nobody believes in it (and conversely a theory which everybody believes in can be false), we rather avoid any talking in terms of beliefs.

As suggested before, the different phases of the problem of induction are presented in decreasing order of importance. Popper considers the 'problem of tomorrow' (the third phase) as less fundamental and less urgent than the previous two. This problem can be formulated as 'How do you know that the future will be like the past?', or to put the question in the language of science: 'How do you know that the laws of nature will continue to hold tomorrow?' Popper's answer to both questions is direct and clear: 'We do not know.' That is, we do not know whether the future will be like the past, and in fact we expect that the future (in some important respects) will be different from the past. We do not know whether what we consider to be laws of nature will continue to hold tomorrow, and we should be prepared for a future possible state of affairs in which some of these laws cease to hold, and under this circumstance we should be ready to explain the situation both before and after such a drastic change.

Popper recognizes that there are many different problems embedded in the foregoing questions. For example, one might mean by the first question that people usually act as if many trends of the past are repeated in the future, and by the second question that if something is a law of nature then it could not possibly cease to hold, because laws of nature are supposed to be completely independent from time-points and space coordinates. Further, there is sharp distinction between the practical side and the theoretical side from which these questions could be asked. Popper acknowledges the difference between practical needs and theoretical reasons, and is ready to consider the appropriate answer. For example, he concedes that for practical reasons one may believe (and expect) that the laws of nature will continue to hold tomorrow, but that for theoretical purposes one should be sceptical. In fact, from the latter point of view, a law-like statement has to be treated as a provisional conjecture. If after many attempts we fail to refute it and it has useful everyday applications, there is nothing wrong with taking advantage of its high degree of corroboration and acting as though it were a law of nature. But one should be prepared to deal with a state of affairs that is incompatible with that type of statement. This can be put in very simple words: one may confidently bet (a practical action) that the sun will rise tomorrow, but should refrain from betting that a certain set of laws – say Newton's laws – will continue to hold in the future, irrespective of any possible criticism.

This recommendation may seem odd. To grasp its import better, the believer in the immutability of law statements must realize that in any dispute about the invariant character of laws of nature there are two issues at stake. The first concerns the customary usage of the expression 'law of nature'. It has been accepted that genuine laws of nature (i.e. statements that express invariant phenomena) cannot cease to operate (being liable to change) and still be called properly 'laws'. Popper does not have any qualms regarding this view. The second issue is different. It has to do with the status of scientific hypotheses and principles. What we call a 'law of nature' is just a statement that refers to some kind of phenomena which remains invariant through a series of changes. If what the statement refers to are changes failing to meet this requirement, we have to accept that this overthrows the purported law. We have to declare that the statement in question was false, and hence that we were wrong. Such a result would be problematic for anybody who considers that linguistic entities like law statements are true, but it poses no problem for anyone who considers statements of this kind as tentative conjectures, since they may be refuted by experience – to nobody's surprise.

As already mentioned, Popper considers that the fourth phase of the problem of induction (which is an aspect of the problem of tomorrow) is metaphysical. He formulates it as follows:

> *There are true natural laws.* This we know, and we know it from experience. Hume says we do not; yet, in spite of what he says, we do: our belief that there are true natural laws is undoubtedly based, in some way or other, on observed regularities: on the change of day and night, the change of the seasons, and on similar experiences. Thus Hume must be wrong. Can you show *why* he is wrong? If not, you have not solved your problem.[26]

Popper shows that this phase is metaphysical in two ways. First, constructing the claim that there are laws of nature by means of the three following singular existential statements: (1) there is at least one true universal statement that describes invariable regularities of nature; (2) some possible universal statement describing invariable regularities of nature (whether yet expressed or not) is true; (3) there exist regularities in nature (whether ever expressed, or expressible, or not).[27] Notice that none of these statements (being *existential*) accurately conveys Hume's challenge, which was directed against the derivability of a *universal* law from a *finite* number of observations. Moreover, notice also that the statement 'there are true natural laws' does not point to any specific natural law, but merely asserts that there is at least one that is true: an assertion that Popper does not find very controversial although he doubts that anybody is in the position of indicating a current formulation that satisfies the requirements for something to be a law of nature along the lines of (1), and (mainly), that it is possible to single it out under the proviso for absolute truth. On the other hand, although the issues connected with (1–3) are quite different, the statements can be treated in the same way, since they have the same logical form (they are existential). On Popper's theory, this amounts to considering them as non-falsifiable metaphysical statements because they in general are not testable (they are neither verifiable nor falsifiable).

Second, neither (1) nor (2) make any reference whatsoever to physics. They both make an assertion about the status of some of the statements that comprise a theory about the natural world, and in this sense they are metalinguistic assertions, though one may accept that they can be interpreted also as advancing a conjecture about the world: 'To assert that there exists a true law of nature may be interpreted to mean that the world is not completely chaotic but has certain structural regularities "built-in", as it were.'[28] However, asserting that there are structural regularities in nature is nothing else than making a metaphysical assertion: one that has little bearing on Hume's argument.

On the other hand, one might confront those who prefer the formulation of the problem of induction in terms of the past–future framework, pointing out that they implicitly take as true a theory of time according to which different moments of time are indeed homogeneous and flow in complete independence of what happens in time (including any possible change in the laws of nature!), just as Newton thought. It is by no means clear that one can make sense of the idea of a future time when there will be regularities different from those acting until now; but even if this is possible, the challengers have to suppose, at least, that the laws of their theory of absolute time will not change and are exempt from Hume's criticism.

As the reader can see, Popper gives his best shot on the problem of induction in the passages I have just commented on. Those who are looking for radically new arguments, however, will be disappointed, since he keeps insisting that one should be able to see the solution to this problem by recognizing that scientific knowledge consists of guesses or conjectures, which allows us to search for truth without assuming a principle of induction, or putting any limits to empiricism. It is not surprising that Popper's solution to the problem of induction does not satisfy many readers, and particularly those who expect a solution in terms of a justificatory principle without the drawbacks of a supposedly discredited principle of induction.[29] He has not found a principle of induction that permits us to derive universal laws from singular statements, nor has he encouraged the idea that it is possible to find a way of establishing the absolute truth of universal theories. He has given an alternative answer to one of the problems encompassed in the general formulation, and has shown that if we accept his solution we would be able to dissolve the challenges of irrationalism and do without induction.

2.3 Criticisms of Popper's solution

Criticisms of Popper's solution to the problem of induction range from the plain denial that a solution has been achieved[30] to qualms about the possibility of obtaining scientific knowledge without induction nor probability. The reader must bear in mind that Popper embraces the following claims: there is a difference between science and pseudo-science, given that there is a method of the rationality of science; there can be no epistemic justification for scientific claims at all, yet there are scientific truths. Since there is no broadly accepted account of epistemic justification available, let alone justification with respect to the sciences, Popper's position seems to be correct. Moreover, he retains the rationality of science and makes the search for truth the aim of science (something with which many agree), but there is a certain cost to be paid and that is what his critics have not fully appreciated. In this section I shall discuss Howson's rebuttal of the Popperian solution and deal with a small sample of criticisms advanced by some authors whom I consider to be representative of the general views in the literature and at the same time have objections clearly articulated. After doing this, I shall try to respond to what I take to be the main critical points raised in this debate.

Howson, to whom I referred briefly at the beginning of this chapter, maintains that Popper's solution to the problem of induction is no solution at all. Howson summarizes Popper's viewpoint with the help of the notion of epistemological neutrality; that is, the idea that, prior to any empirical investigation, one should assign zero probability to any universal hypothesis which is not equivalent to a singular statement (interpreted in a universe of infinite size), if one wishes to avoid any bias in favour of a particular type of universe. The problem is that Popperians get trapped in the following inconsistency: (1) all universal hypotheses have zero inductive probability (inductive probability is defined as the probability of a hypothesis being true given any (finite) number of confirming evidential statements) and (2) when confronted by a refuted and a not-yet-refuted (call it unrefuted) hypothesis, one has reasons to prefer the latter. Howson's argument runs like this: suppose there are two alternative hypotheses h_1 and h_2, where h_1 is refuted and h_2 is not yet. Since $p(h_1 = t) = p(h_2 = t) = 0$, then $p(h_1 = f) = p(h_2 = f) = 1$ (where 't' means true and 'f' false). Also since the two hypotheses are the only two alternatives (considered), $1 = p(h_1 \text{ and } h_2 = f) = p(h_1 = f) \cdot p(h_2 = f)$, namely they are independent. Even though there seems to be the difference that h_1 is known to be false

Solution to the Problem of Induction

whereas h_2 is not, which may be regarded (a move apparently available to Popper) as a difference enough to prefer h_2 to h_1, the difference is not useful to Popper. Although h_2 may still turn out to be true while h_1 is known to be false, as far as preference of hypotheses, which should be based on $p(h=t)$, for any h, h_1 and h_2 are equally likely to be false. Therefore, Popper is forced to be epistemically neutral to these two hypotheses.

There is a quick reply to this argument. When choosing between competing hypotheses and expressing preference for the unrefuted over the refuted, one has no commitment to the truth of the unrefuted hypothesis. Popper says only that the unrefuted hypothesis *may* be true (whilst the refuted hypothesis *must* be false), and this alone suffices to make it a better alternative. Before dismissing this reply, Howson restates it in terms of what the decision-theorists call a dominance argument. If h_1 has been refuted in a crucial experiment between it and h_2, then although one cannot assert that h_2 is true, one can see that it dominates over h_1 (which *cannot* be true); hence, it is preferable. However, Howson thinks, this does not do it. Where the matter of the preference is *truth*, there are no reasons for strictly preferring h_2 over h_1. Moreover, he considers that in the present example Popper is inferring the reasonableness of a preference for h_2, from the unreasonableness of a preference for h_1 – a fallacious move. On the contrary, the situation is such that it is possible to remain reasonably indifferent between the competing hypotheses:

> Suppose the alternatives to h_1 were uncountably infinite in number, each having, with h_1, probability zero. The situation we contemplate is now isomorphic to an uncountable fair lottery, in which we are given the choice of two tickets, one of which we know from an oracle is not the winning ticket, and about the status of the other we know nothing at all. The expected value of both tickets is the same, i.e., 0, and consequently in such a situation we ought, I claim, to be perfectly indifferent between them.[31]

In other words, Howson argues that Popper's requirement of epistemological neutrality harms his own position: given that the probability of all universal hypotheses is zero and that there are uncountable alternatives to every scientific hypothesis, the epistemologist who is asked to choose from a set (whether finite or infinite) of competing hypotheses (only one of which has been falsified) has no good reason to prefer the unrefuted hypothesis as a suitable candidate for the truth (disregarding the qualifications – perhaps better – of many other possible candidates). Howson considers that his model of the infinite lottery is adequate to appraise the rationality of choice between competing hypotheses because it parallels a common situation in science, one where we are presented with two functions only one of which is in the class that has been ruled out by the measurements, and we are asked to choose between them under the understanding that the remainder (unknown) functions are uncountably infinite in number. Again, when there are infinite mutually exclusive candidates for truth, a preference for the unrefuted hypothesis over the infinite number of competitors cannot reflect a preference for truth over falsity, and even a Popperian should be able to see this if he takes into account that all those alternatives have equal probability, namely zero. The upshot of Howson's argument is that Popper's anti-inductivism commits him to a thoroughgoing scepticism in science. His positive conclusion is that any reformulation of the problem of induction has to be very weak in order to admit any constructive solution, and that Popper's reformulation is still too strong for such a result.

I find Howson's arguments both mistaken and unconvincing. To begin with, he claims that Popper remains silent in relation to what a preference for truth entails in the way of practical actions and decisions involving the choice between competing hypotheses. But this is not the case, since Popper has repeatedly advised us to choose the unrefuted and best-corroborated hypothesis. I think Howson is not satisfied with this advice because he neglects to note that in some contexts Popper replaces the search for truth by the search for verisimilitude. As we will see in Chapter 5 below, when confronted with two competing hypotheses one does not know if the unrefuted hypothesis is true, but one does know that it *may* be true, and provided it is well corroborated, one has good reasons to believe that it is closer to the truth than the falsified competitor. Now, Popper explicitly emphasizes that the reasons that support this belief are not justificatory reasons and might not be, strictly speaking, rational. However, despite such possibility, they are the best reasons available for effects of practical action, making epistemic preference for the unrefuted hypothesis the natural result.[32]

On the other hand, I do not think that Howson's model of the infinite lottery provides a good representation of the situation he is examining. Assuming that the lottery is fair, all tickets have the same probability to win. Therefore, if I were a gambler wanting to buy a ticket (a practical action) and were confronted with the situation Howson's example portrays, I see no reason to remain perfectly indifferent between the two tickets. I would certainly want to buy the ticket that has not been ruled out as a possible winner. To buy the other would be simply stupid. Now, if the point is that only one among an infinite number of tickets is the winner and knowing that one is not does not lead us any closer to picking *the* winning ticket, then he is right. But I think this point misrepresents the meaning of Popper's answer to the question of what criteria one should use to select between competing hypotheses.

Although Howson claims that the subjective–objective distinction has no bearing on his lottery-model argument (granting the epistemic neutrality of his ideal observer), he pictures a situation in which a subject is presented with

Solution to the Problem of Induction

two alternatives and is asked to show his preference. It is not very difficult to show that Howson's analogy is far from perfect. Call the discarded ticket d and its competitor e. For Howson's argument to work he needs to establish a parallel between d and h_1, on the one hand and e and h_2, on the other. Let us see if this parallel holds. We can represent the situation schematically like this (I am letting the special character '}' stand for 'is asked to choose'):

$$G \; \} \; d \text{ or } e$$

$$S \; \} \; h_1 \text{ or } h_2$$

According to the conditions of the example, we have two cases in which the subject (the gambler or the scientist) has information only about the first component of the alternation, and knows nothing about the second component. Well, perhaps this is too strong a way to put it; let us say that the gambler and the scientist, respectively, know that their second option is just one among an infinite number of equiprobable alternatives. Thus, neither of them has any grounds whatsoever to make a reasonable choice. Consequently, they should remain perfectly indifferent between both options. Let us assume that my previous critical remark concerning the gambler who wants to buy a ticket is misconceived, for it is insensitive to the fact that (prior to the drawing) each of the members of the infinite ticket set has probability zero. Should we agree with Howson's criticism of Popper? I do not think so. There are at least two reasons that make his example inadequate.

We can see easily that the parallel does not hold, by noticing that e and h_2 are not comparable. Whereas the gambler does not know anything about e, except that it has not been ruled out by the oracle (unlike d), the scientist does have a crucial piece of information about h_2, one that can provide rational grounds (though not necessarily a *justification*) to support his preference for it. He knows that h_2 is well corroborated, and presumably that it has survived the crucial experiment designed for both h_1 and h_2, and this alone may, in my opinion, change the situation drastically. Why should the scientist refrain from expressing his preference for h_2 over h_1? Howson contends that this is because h_2 is only one among an infinite number of mutually excluding competing hypotheses, each of which has probability zero. Although, technically speaking, this is correct, and has been recognized by Popper, I think it is improperly called into play in this particular instance. Except in cases of random choice, in order to select one among a group of competing hypotheses, one must know which are the members of the set. So even in the case in which *theoretically* the number of competing hypotheses is infinite, one cannot consider the overwhelming majority of completely unknown potential competitors to make the choice; one chooses between the two that are presented for the selection in the first place. Again, regardless of the infinite number of potential competing hypotheses, scientists are hardly ever confronted with more than just a few of them at a time, so it seems perfectly rational to prefer (if choosing only between two) the hypothesis that has survived falsification and is well corroborated.[33] I can see why Howson finds this view unpalatable: he stresses that when the criterion which guides choice between competing hypotheses is *truth*, we have no grounds to bestow attributions of truth upon h_2, and this is a fair remark. But he also knows that in the case at hand we are strictly concerned with possible truth or actual verisimilitude, and the latter gives us reasons to express our preference. Furthermore, Howson was complaining about Popper's failure to indicate what a preference for truth entails in matters of practical action, but he is guilty of a more serious crime. Though he structures his objection as a putative case of practical action, he uses arguments and epistemological features (i.e. the infinite number of competing hypotheses) that are compelling exclusively in theoretical matters. No wonder he considers Popper's suggestions wanting.

Before bringing this chapter to an end, I will summarize Miller's formulation and elimination of the most common objections against the stability of Popper's anti-inductivist stance.[34] I will expound briefly five of such objections, while deferring most of my comments on corroboration as a disguised form of induction until Chapter 4 below.

The first and most obvious objection states that in order to obtain scientific knowledge one has to presuppose at least some regularities; in other words, that scientific method can only be applied if scientists assume that natural processes are immutable. This point is made, among others, by O'Hear who deems Popper's attempt to dispense with induction unsuccessful, since inductive reasoning 'removed from one part of the picture, crops up in another ... the underlying reason for this is that any coherent conceptualization of experience requires the assumption of a stable order of the world'.[35] Nowhere has Popper (or Hume) denied the existence of regularities in nature, nor has he suggested that they should be unknowable, nor that genuine laws can be refuted in the sense that the regularities described by them may arbitrarily change, since it is a part of the meaning of the expression 'scientific law' that natural laws are to be invariant with respect to space and time. But this is a methodological rule that can be distinguished easily from a rule of inference like the rule of induction.[36] In fact, Popper anticipated this criticism in *LScD* where he explained that the 'principle of the uniformity of nature' was, indeed, a metaphysical principle, often confused with an alleged inductive rule of inference, precisely because the principle of induction is also metaphysical. Moreover, he acknowledges that the metaphysical faith in the principle of uniformity of nature is required for practical action, and contends that the corresponding methodological rule (expressed very superficially by such a principle) requiring that any new system of hypotheses yields the old, corroborated regularities can be derived from the nonverifiability of theories.

Consequently, he carefully distinguished the metaphysical principle from the methodological rule, for falsificationism is not committed to the existence of regularities in nature. To make this point more clear, let us emphasize that in order to provide interesting knowledge about the world, the inductivist needs to assume not only that all or most apparent regularities are genuine laws (a false and non-falsifiable principle), but also that *there are* regularities in the world. By contrast, the falsificationist only requires that some regularities operate in the world, without having to make any assumption to this effect, that is without any ontological commitment to the regularities. From his freer standpoint, Popper concludes that:

> ... from a methodological point of view, the possibility of falsifying a corroborated law is by no means without significance. It helps us to find out what we demand and expect from natural laws. And the 'principle of the uniformity of nature' can again be regarded as a metaphysical interpretation of a methodological rule – like its near relative, the 'law of causality'.[37]

A second type of objection targets the two main possible outcomes of the process of theory-testing. Roughly speaking, it charges that a corroborated hypothesis should pass similar tests in the future; and correspondingly, that one who discards a falsified hypothesis is presupposing that it will fail future tests again. Either presupposition, the critics claim, brings induction back into the picture. The first variant of this objection was raised by Levison who thought that Popper's account of science could not eliminate the problem of induction, or do without some sort of verification. In his opinion, favourable testing of a theory can provide us with a reason for preferring one of its predictive consequences to any of its possible contraries only if we allow that past instances of successful prediction that we have observed will be equally successful in future (unobserved) instances: 'even if the prediction is specifically identical to predictions that have been endlessly verified in the past. Some concept of inductive reasoning, or extrapolation, is needed, therefore, in order to justify supposing that an experiment can be successfully repeated.'[38]

Ayer has formulated the second variant in different ways. Although he admits that relying on the positive outcome of a test – based in some favourable instance in the past – may be an unwarranted assumption, he wonders why should we make every effort to discover counterexamples to a given hypothesis if it gains no credit from passing the test.[39] In a similar vein he asks elsewhere: 'Why should a hypothesis which has failed the test be discarded unless this shows it to be unreliable; that is, except on the assumption that having failed once it is likely to fail again?' Apparently, Ayer never considers the possibility of rejecting a theory simply because it is false, since he feels that falsificationism is unable to give a satisfactory answer to his questions. In the same fragment just quoted he adds:

It is true that there would be a contradiction in holding both that a hypothesis had been falsified and that it was universally valid; but there would be no contradiction in holding that a hypothesis which had been falsified was the more likely to hold good in future cases. Falsification might be regarded as a sort of infantile disease which even the healthiest hypotheses could be depended on to catch. Once they have had it there would be a smaller chance of their catching it again.[40]

Ayer's position is surprising since Popper never makes or commends any inference from past falsification to future falsification. On the contrary, he says that falsification (as is the case with the whole game of empirical science) should be open to critical examination and that, all in all, no falsification is ever definitive.[41] Thus, the possibility that Ayer is contemplating (that today's falsified hypothesis may become tomorrow's corroborated) is not extraneous to Popper's theory of science, and is not ruled out of the actual world, for Popper does not say that a hypothesis that fails a test will necessarily fail any conceivable repetition of that test, although (normally) failure means that the hypothesis is false. Of course, one should rule out the stratagem of the conventionalist who wishes to reinstate a falsified hypothesis, either by denying that it has failed the test or by introducing an *ad hoc* modification to save it from falsification, but this is another story. In sum, performance on a series of tests (regardless of their outcome) is not governed (and *a fortiori* not *justifiable*) by a purported principle of induction. So much for the objection concerning the stability of a hypothesis's performance on various tests.

The third objection is related also to the repeatability of tests, focusing on the issue of their severity. Popper held the view that once a theory passed a severe test, additional tests of the same kind increase its degree of corroboration very little (unless, of course, the new instances were very different from the earlier ones).[42] And in the *Postscript*, where he discusses the notion of corroboration at length and shows that it should not be confused with verification or treated in terms of probability, he argues that there is something like a law of diminishing returns that prevents us from ascribing the same (decisive) value to further iterations of a severe test (whether passed or failed by the theory). Thus Popper discussed the problems of the existence of severe tests and whether their severity declines with repetition under the assumption that one can do science without induction. Several authors have resisted both Popper's conclusion and his assumption. For example, O'Hear denies that there could be a law of diminishing returns which is independent of induction. He claims that anti-inductivists lack reasons to think that a theory that has survived a specific test is likely to survive similar tests on the n+1 occasion they undergo them: 'For what could be the basis of such a view other than a generalization from instances of which we have had experience to

Solution to the Problem of Induction 31

one – a different one, even if in some ways similar – of which we have as yet had no experience?'[43]

And a few lines below he says that on any reading, background knowledge appears to be used inductively in determining test severity. There is nothing to be said regarding the first part of this qualm. But in *CR*, Popper makes clear that assessments of the severity of tests are relativized to background knowledge instead of past experience, and, hence, no inductive assumption is required. Besides reports of past performance in the testing ordeal, background knowledge contains also generalizations of many kinds, such as, for example, the standing of a theory in relation to its competitors and the state of the problem situation. Therefore, only a severe test can help us to exclude inadequate assumptions or prejudices that may have sneaked into our background knowledge. On the other hand, the results of testing are either falsification or corroboration. If the former, we should not worry any more about the severity of the tests (provided they have been carefully conducted), since the theory has been (presumably) falsified; if the latter we should not expect, under the assumption that we are employing a reproducible effect, that the likelihood of the theory to survive the same test again has increased – as O'Hear rightly observes it. In fact, the requirement to subject theories to severe tests is a methodological requirement (not an epistemological one), and Popper points out that as we impose severe tests upon a theory, we are left with fewer new tests to which we can expose it. Yet this does not sound like an inductive use of background knowledge.

The fourth type of objection charges that Popper's falsificationism is troubled by Goodman's paradox, just as inductive theories are. This paradox is so widely known that I shall not provide a detailed account of it. Let me just mention that Goodman shows that the same evidence that confirms the hypothesis 'all emeralds are green' confirms 'all emeralds are grue', where Goodman's coined predicate means 'green if and only if first observed before the year 2000; blue otherwise'. The charge seems to be that there is no solution to Goodman's new problem of induction without introducing an appropriate distinction between sound and unsound inductive inferences; that is, without presupposing an inductive principle.[44] The falsificationist – the objection proceeds – has no empirical reasons to prefer one hypothesis over the other, so he can neither endorse the prediction that emeralds observed after 1 January 2000 will be green, nor that they will be grue, and since these predictions are mutually incompatible, he will not be able to use the method he has highly commended. Popper referred to this objection in the introduction to the first volume of his *Postscript*. He answered it in the following way:

> That [my theory can answer Nelson Goodman's paradox] will be seen from the following considerations which show, by a simple calculation, that the

evidence statement e, 'all emeralds observed before the 1st of January of the year 2000 are green' does not make the hypothesis h_1, 'all emeralds are green, at least until February 2000' more probable than the hypothesis 'all emeralds are blue, for ever and ever, with the exception of those that were observed before the year 2000 which are green'. This is not a paradox, to be formulated and dissolved by linguistic investigations, but it is a demonstrable theorem of the calculus of probability.[45]

The theorem to which Popper is alluding asserts that the calculus of probability is incompatible with the conjecture that probability is ampliative and therefore inductive. The main reason is that increasing favourable evidence does not change the ratio of the prior probabilities calculated (or assigned) to any pair of competing (and incompatible) hypotheses so that they both can explain the same evidence. But one could also respond that 'all emeralds are grue' was compatible with any possible test conducted before the deadline, and therefore should not be admitted to empirical science, and *a fortiori*, that it should not be admitted as a genuine competitor of 'all emeralds are green', since no test can adjudicate its claim to be true. As for the charge that falsificationists lack grounds to prefer any of the competing hypotheses because they cannot rationally justify their choice, one can say two things. First a choice could be rational even in the absence of justification (recall that Popper's project is not justificationist). Second, the predicate 'grue' does not solve any problem that is not equally well solved by its natural counterpart 'green', in fact, we can say that it does not solve any problem at all (for this reason, as Miller points out, gemologists are not concerned about it); hence, it need not be admitted in empirical science.

The last sort of objection is related with the context of practice. The critics start by taking for granted that empirical evidence provides good support for the success of technology (pragmatic science). But, they claim, this is just a form of induction. Hence, Popper has to reject technology or accept inductive evidence into his theory. To put the argument in other form, unless past observed instances of successful prediction based on a theory give us reason to suppose that future instances will be similarly successful, we have no reason to conjecture that a prediction that follows from a well-tested theory may be true rather than false, and, *a fortiori*, no reason for preferring a particular prediction to one of its rivals, and this is so even if the prediction is identical to one that has been endlessly verified in the past. We cannot make the choice without supposing some principle of induction or extrapolation to justify that a prediction can be successfully tested again. For example, Cohen argues that although Popper's philosophy has excelled in developing Hume's lesson, it has produced a kind of science which is quite unusable. His theory has been unable to depict a plausible form of rationality that can be attributed to technology, i.e., to the selection of ... [those] scientific hypotheses that are to be relied on for our prediction about the behaviour of physical forces or substances in particular cases. You would be rather rash to be the first person to fly in a plane that had been made in accordance with the boldest conjectures to survive ground tests on its materials, if extreme conditions covered by the bold conjectures were not present in any of the tests. The conjectures would have relatively high Popperian corroboration but be ill-supported by the evidence. The reasons for accepting them would be rather poor.[46]

And in the same vein other critics argue that it is not possible to make rational decisions on how to act without appealing to some sort of inductive principle. Moreover, they think that any degree of confidence in technological devices, as well as any assumption on the invariance of scientific laws, is possible only if one subscribes to a principle of induction. But, as discussed above, this is a mistake. One who believes that the laws of physics that hold today will continue to hold tomorrow, need not subscribe to any principle of induction. His belief concerns only the laws of physics and likewise the prediction that they will hold tomorrow. Still others[47] have charged that a Popperian would lack reasons to choose between competing theories (even if one has been refuted) for practical purposes. In *OK*, Popper specifically addresses this objection. To the question about which theory we should *rely* on for practical action (from a rational point of view) he answers that – given that no theory has been shown to be true, or can be shown to be true – we should not rely on any theory. To the question which theory we should *prefer* for practical action (again, from a rational point of view), he says that we should prefer the best-tested theory:

> In other words, there is no 'absolute reliance'; but since we have to choose, it will be 'rational' to choose the best-tested theory. This will be 'rational' in the most obvious sense of the word known to me: the best-tested theory is the one which, in the light of our critical discussion, appears to be the best so far, and I do not know of anything more 'rational' than a well-conducted critical discussion.[48]

Notice that Popper suggests that the rational agent should act as if the best-tested theory were true, but he only invokes the state of the critical discussion. Further, he does not justify why we should act that way (nor does he imply that the falsificationist has a *justification*), he remains content with pointing out that acting that way is the most rational choice under the circumstances. On the other hand, if one has to choose between a refuted and an unrefuted theory, the obvious conjecture is that the theory that has best survived testing becomes the best source of true information about the world, and it would make no sense to act as if this theory were not true. As for Cohen's worries, it seems that he is conflating two different problems. He is right when noting that Popper's theory of science gives prescriptions applicable mainly to general theories. But then he claims that a Popperian is unable to choose rationally the theory that will make the best predictions, and he charges that he would not have good reasons to trust a device that has been constructed in accordance with falsificationism. Both charges are inappropriate. The former because – as seen before – there is no difficulty involved in the rational choice of the best theory, and this includes the theory that yields the best predictions. The latter because there is no scientific theory that can tell anybody what is the right way to construct artifacts – planes or what you like – and if the test of their materials were conducted in the absence of extreme conditions, even inductivists would be in great danger. But perhaps the main motivation for disagreeing with Cohen's charge is that Popper's theory of science is not concerned with the problems that arise in the construction of devices but with the interesting problems of scientific theorizing.

Though this review on the most common criticisms against Popper's stance on the problem of induction is by no means exhaustive, I think it gives the reader a good idea of what is going on in the literature. Now that we have sufficient background about the motivation for a non-inductivist philosophy of science, we are ready to move on to its key notion.

Notes

1. I do not want to suggest that Popper's claim of having solved the problem of induction was completely unknown before this time. I hold that a more articulated discussion of it was brought about by the publication of *LScD*. For a similar opinion see Fain's review of *LScD* in: Fain 1961, pp. 319ff.
2. For example, Warnock, in his review of *LScD*, contends that Popper recognizes the insolubility of this problem and that his arguments cannot eliminate it; hence 'if there is such a thing as the problem of induction, it is a problem for Popper as much as for anyone else' (Warnock 1960). Warnock's reluctance to admit Popper's success is motivated mainly by his inductive supposition that a theory which has withstood rigorous tests in the past 'proves its fitness to survive' in future tests (pp. 100–1). Warnock seems to be wrong on two counts: (1) given that Popper offers a solution to the problem of induction – though not a justificatory one – he could not possibly have said that the problem has no solution; (2) the reviewer assumes Popper meant that a theory that has survived until now, and that has thereby proved its fitness to survive until now, has also proved its fitness to survive future tests, but this is contrary to what Popper maintains.

3. Incidentally, even philosophers who are friendly to Popperian ideas take for granted that the Problem of Induction has not been solved. One can read in this spirit Watkins' recent article 'How I almost solved the problem of induction', (Watkins 1995).
4. 'The contrary of every matter of fact is still possible, because it can never imply a contradiction' (Hume 1993, section IV, part I, p. 15).
5. On the other hand, on Hume's view any argument that appeals to experience in order to ground our confidence in the similarities between past and future – an aspect essential to inductive inferences – is somewhat circular. He writes on the same page of his sceptical remark about this relationship: 'it is impossible, therefore, that any arguments from experience can prove this resemblance of the past to the future; since all these arguments are founded on the supposition of that resemblance' (Hume 1993, section V, part II, p. 24).
6. The basic problem is that there are only two ways of justifying inferences from past to future experience. Either one does it by appealing to a-priori principles or by appealing to contingent ones. In either case the justification would be circular. Since this problem should be familiar to the reader, I assume this brief explanation will suffice to make sense of the forthcoming discussion.
7. Hume 1993, section V, part I, p. 28.
8. Cf. ibid., section VII, part II, p. 51.
9. 'All reasonings concerning matter of fact seem to be founded on the relation of *Cause and Effect*. By means of that relation alone we can go beyond the evidence of our memory and senses' (Hume 1993, section IV, part I, p. 16). Moreover, our reasoning about factual questions supposes a connection between a past fact and another that is inferred from it. The cause–effect relation enables us to draw inferences that go beyond memory traces and sense impressions and becomes the foundation of science as well as of our knowledge about the existence of objects.
10. Hume 1993, sections V and VII, *passim*.
11. Hume 1978, Book I, part 3, section VIII, p. 103.
12. Hume suggested that his sceptical conclusions did not affect ordinary life, nor did facts of ordinary life refute his conclusions. He pointed out:

> my practice, you say, refutes my doubts. But you mistake the purport of my question. As an agent, I am quite satisfied in the point; but as a philosopher, who has some share of curiosity, I will not say scepticism, I want to learn the foundation of this inference. No reading, no enquiry has yet been able to remove my difficulty, or give me satisfaction in a matter of such importance. (Hume 1993, section IV, part II, p. 24)

13. *LScD*, p. 28; cf. also *CR*, pp. 21–5.
14. *LScD*, p. 29.
15. Ibid., p. 33.
16. *CR*, p. 54 (parentheses suppressed).
17. *OK*, p. 4.
18. Ibid., p. 7.

19. The idea that theories are truthlike instead of true is problematic. I will discuss the issue in Chapter 5.
20. Cf. *OK*, p. 12.
21. 'How can we adjudicate or evaluate the far-reaching claims of competing theories or beliefs?', *RAS*, p. 19.
22. A few pages later Popper writes:

    > What I have said here provides a complete solution to Hume's logical problem of induction. The key to this solution is the recognition that our theories, even the most important ones, and even those which are actually true, always remain guesses or conjectures. If they are true in fact, we cannot know this fact; neither from experiment, nor from any other source. (Ibid., p. 33)

23. Ibid., p. 23.
24. Ibid., p. 35.
25. More clearly expressed: 'Hume's argument does *not* establish that *we may not draw any inference from observation to theory:* it merely establishes that we may not draw *verifying* inferences from observations to theories, leaving open the possibility that we may draw *falsifying* inferences' (ibid., p. 54).
26. Ibid., pp. 71–2.
27. Cf. Ibid., p. 72.
28. Ibid., p. 74

29. It seems to me that all the objections to my theory which I know of approach it with the question of whether my theory has solved the traditional problem of induction – that is, whether I have justified inductive inference.
    Of course I have not. From this my critics deduce that I have failed to solve Hume's problem of induction. (*OK*, p. 28)

30. Among the many that raise this criticism it is worth mentioning Maxwell, who thinks that in so far as Popper's attempted disposal of induction is based on the falsifiability requirement, it seems to fail (Maxwell 1974) and Levison, who argues that Popper did not succeed in solving Hume's problem (Levison 1974).
31. Howson 1984, p. 144.
32. Howson's argument is based on an unstated assumption that one should use probability to choose between competing hypotheses. But probability is not the only criterion one could use to select a hypothesis, and obviously Popper is not bound to that assumption (that turns out to be false). The reader may consider, in addition, that in Popper's theory expressing a preference for a competing hypothesis does not involve an assessment of probabilities; hence, it will not be jeopardized by the fact that all competing hypotheses have the same numerical probability. Preference is always a matter of advancing conjectures about how close to the truth a certain hypothesis may be. As it happens with all conjectures, it may turn out that we choose wrongly, but this is just one of the risks of the Popperian game.
33. On Howson's view, restricting the number of competing hypotheses to the ones actually presented before the scientist provides a trivial solution to the difficulty and furnishes a very limited defence against the sceptical questioner of

scientific procedures. Moreover, since this move cannot establish the superior reasonableness of preferring an unrefuted to a refuted hypothesis, it makes the problem of rational choice within the class of unrefuted hypotheses devoid of interest. (Cf. Howson 1984, p. 145.)

34. The best treatment of the criticisms of Popper's views on induction is due to David Miller. I refer the interested reader to Miller 1994, Chapter 2. This is a slight re-elaboration of Miller's paper 'Conjectural knowledge: Popper's solution of the problem of induction' which appeared in Levinson 1982, pp. 17–49. I agree with most of Miller's treatment of this controversy, and my own views are greatly influenced by his arguments.

35. O'Hear 1980, pp. 57–8. I should add that Popper agrees with the assumption of invariance, but he contends this is a metaphysical assumption, whereas induction is an illusion.

36. Cf. *LScD*, p. 253; *RAS*, pp. 66; 71–78.

37. *LScD*, p. 253.

38. Levison 1974, p. 328.

39. Ayer 1974, p. 686.

40. Ayer 1956, p. 74. The same point appears in O'Hear 1980, p. 63 as well as in Burke 1983, pp. 59–60, who sides with Ayer.

41. *RAS*, p. xxii; cf. also *LScD*, p. 50.

42. *LScD*, p. 269.

43. O'Hear, 1980, p. 45. See also Musgrave 1975, p. 250

44. The objection, voiced among others by Feyerabend 1968, is very popular. Levison – one of its fans – claims that since Popper's theory failed to solve the problem of induction, there is not much point in finding out whether it also fails to solve Goodman's paradox. However, he thinks that 'it will be possible to show briefly that Goodman's puzzles constitute as much a problem for Popper as for any "inductivist"'. (Levison 1974, p. 323)

45. *RAS*, p. xxxvii.

46. Cohen 1980, p. 492.

47. O'Hear 1975, 1980; Feigl 1981.

48. *OK*, p. 22.

3
Falsifiability

> I hold that scientific theories are never fully justifiable or verifiable, but that they are nevertheless testable. I shall therefore say that the *objectivity* of scientific statements lies in the fact that they can be *inter-subjectively tested*.
>
> (*LScD*)

Falsifiability is a key notion in Popper's theory of science. Not only does it encapsulate the Popperian solution to the problem of induction, it yields his restatement of the problem of demarcation and becomes the leading hallmark of empirical theories.[1] Falsifiability is a property, a sort of potentiality used to characterize any theory that possibly may clash with a particular state of affairs. According to its more simple formulation, it tells us that, when such a clash takes place, the theory in question becomes *falsified*.[2] This terminological point is important since Popper uses the notion of falsification only to refer, in general, to cases of actual clash. There are several definitions of falsifiability available in *LScD*, but they all emphasize the realizability of the feature (I will go shortly to the definitions). Falsifiability comes in degrees, and it is also a way of conveying the 'empirical' or 'informative' content of a theory. The more falsifiable a theory is, the higher its informative content and the better it fits (for the purposes of theory selection) as a part of empirical science. However, according to Popper, all theories are nothing more than conjectures, and so are their predictions – which are, in their turn, falsifiable. But, since falsifiability is a matter of degree, not all predictions have equal status: some are easier to falsify (i.e. riskier or bolder) than others, and it is precisely this characteristic that accounts for their value. Consequently, one who is participating in the practice of science should direct his/her attention exclusively towards theories that make bold conjectures and anticipate risky predictions, because they have the feature – falsifiability – in the most genuine sense. The bolder the conjecture or prediction, the more we discover when the prediction is not falsified.

3.1 Deductivism and falsifiability

I have already surveyed Popper's main objections against induction as a suitable methodology for empirical science. To the dismissal of inductive methods, motivated by the difficulties of giving a justification for reductive inferences, we need to add the denial of the existence of an inductive logic. Regarding the latter, Popper's stance leaves no room to doubt: 'As for inductive logic, I do not believe that it exists. There is, of course, a logic of science, but that is part of an applied deductive logic: the logic of testing theories, or the logic of the growth of knowledge.'[3] It is important to emphasize that Popper regards his theory of science as entirely deductive. After having rejected induction for all the reasons explained in the previous chapter, Popper thinks that it is still possible to account for the growth of scientific knowledge without incurring the problems that the inductivist faced. In other words, Popper contends that there is a way out of Humean inductive scepticism.

As is well known, Popper is interested in the epistemological problems pertaining to empirical science. These problems cover a wide range of topics, and I will not examine all of them here. Nevertheless, to give an idea of the fruitfulness of the notion of falsifiability, let me just mention that it plays a decisive role in two of the most important activities in science: namely, the processes of theory-testing and theory selection. More importantly, on his account, the processes of theory-testing can be carried out deductively according to the inferential rule of *modus tollendo tollens*, which by the falsity of the conclusion enables us to reject the premises with deductive certainty. The process works as follows:

> With the help of other statements, previously accepted, certain singular statements – 'which we may call 'predictions' – are deduced from the theory; especially predictions that are easily testable or applicable. From among these statements, those are selected which are not derivable from the current theory, and more especially those which the current theory contradicts. Next we seek a decision as regards these (and other) derived statements by comparing them with the results of practical applications and experiments. If this decision is positive, that is, if the singular conclusions turn out to be acceptable, or *verified*, then the theory has, for the time being, passed its test: we have found no reason to discard it. But if the decision is negative, or in other words, if the conclusions have been *falsified*, then their falsification also falsifies the theory from which they were logically deduced.[4]

Thus the process of testing has two possible outcomes: positive or negative. If the former obtains in the face of detailed and severe tests, and if the theory is not superseded by another theory in the course of scientific progress, then one could say that the theory has 'proved its mettle'; if the latter occurs, the theory becomes falsified.[5] In further work Popper will use exclusively the adjective 'corroborated' (in opposition to 'verified') to refer to a theory that has passed a test. This is, certainly, due to an effort to distinguish his views from the positivist's. One has to keep in mind that a positive result can only give temporary support to a theory, since there is always the possibility of a subsequent failure to withstand another test that can overthrow it.[6]

It should be clear, by this time, that the procedures of theory-testing do not involve inductive logic, let alone make any concession to the popular belief (among inductivists and verificationists) according to which it is possible to argue from the truth of singular statements to the truth of theories, or to establish the truth of theories based upon their verified conclusions. The Popperian can do it with just the laws of logic (which do not require justification because they are analytic) as long as he renounces the requirement that the statements of empirical science must be conclusively decidable, i.e. that it should be logically possible to establish their falsity or truth. In Popper's opinion, a major asset of his deductivist approach to testing, and his contention that unilateral falsifiability suffices for empirical science, is that it can deal with the most interesting epistemological problems and dissolve the problems raised by a purported inductive logic 'without creating new ones in their place'.[7] Deductive testing of theories proceeds in the following way: from a system of statements we deduce statements of lesser universality. Likewise, these statements are tested with the help of less universal statements, and so on. Popper explains that there is no danger of an infinite regress in this procedure because it does not intend to justify or establish the statements that are being tested, nor does it require that every statement must be actually tested, but only that it be in principle testable.[8]

Popper's philosophy respects the principle of empiricism, and even takes it one more step ahead since his whole project is concerned with *empirical* science. On Popper's view a theoretical system is to be admitted as scientific (or empirical) if it is testable by experience; that is, if its logical form makes empirical refutation possible. In other words, a theoretical system is empirical if, and only if, it is falsifiable. Popper has repeatedly emphasized that this proposal is not liable to the same kind of objection that he raises against the verificationist criterion of demarcation, because it is based on an asymmetry between verifiability and falsifiability. On his view, grasping this asymmetry is crucial to understanding how falsifiability can readily yield what verifiability is utterly unable to do (at least in view of providing a suitable demarcation criterion). This is because universal statements can never be derived from singular ones, though singular statements can contradict them. Therefore, we can argue from the truth of singular statements to the falsity of universal statements using just deductive inferences. 'Such an argument to the falsity of universal statements is the only strictly deductive kind of inference that proceeds, as it were, in the "inductive direction"; that is, from singular to universal statements.'[9]

Falsifiability

An objection to the logical value of Popper's criterion of demarcation points out the difficulties of falsifying conclusively a theoretical system. First, one who wants to evade falsification may use the moves of the conventionalist: introducing *ad hoc* auxiliary hypotheses or changing *ad hoc* a definition. Second, it is always possible without incurring in any logical inconsistency, as Popper remarks, to simply refuse to acknowledge any falsifying experience. The former stratagem can be blocked by restricting the introduction of auxiliary hypotheses and the modification of definitions: new hypotheses and definitions are welcome only if they do not decrease the falsifiability of a theoretical system; that is, if they do not attempt to 'save the lives of untenable systems'. The latter should be disavowed as an instance of the wrong attitude in empirical science: one could refuse to accept a falsifying experience only through critical discussion and examination of the problem situation.

Popper referred again to the asymmetry between verifiability and falsifiability in the *Postscript*. It is undeniable that a set of singular observation statements *may* falsify a universal statement, whereas there is no possible situation in which such a set *can* verify it, in the sense of establishing or justifying it (conversely, such a set can verify an existential statement but cannot falsify a purely existential statement). For this reason, Popper usually writes that universal statements are 'unilaterally falsifiable', whereas existential statements are 'unilaterally verifiable', and holds that these two ways of deciding upon the truth-value of statements serve different purposes in epistemology. But the critics do not give up easily. They rejoin that the falsification of a statement automatically presupposes the verification of its negation; hence any talk in terms of verification or falsification turns out to be based on a mere verbal difference, because they are completely (logically) symmetrical. Although Popper does not mention it specifically, he must have in mind Carl Hempel's version of falsifiability in his report about the changes in the positivist criterion of verification.[10] The line of thinking Hempel is summarizing is grounded in a property and a desideratum of statements: (a) the logical equivalence of universal statements and the negation of their existential counterparts, wherein falsification of a statement (or theory) presupposes the verification of its negation and vice versa; and (b) that for purposes of logical classification, both a statement and its negation should be accorded equal status. Given (a) any distinction between verification and falsification becomes pointless; given (b) if a statement is declared meaningful (or meaningless), so must be its negation. This point can be further illustrated as follows: suppose that a falsificationist is trying to locate the observation statement that will refute theory A; then, we can consider that both the difficulties he faces as well as the successes he achieves are the mirror images of what an imaginary positivist rival who is attempting to verify theory $\sim A$, will encounter. *A fortiori*, if he does succeed in falsifying A, there is no reason to deny that he has also verified $\sim A$.

This, however, is misleading. To begin with, we need to conduct this discussion in terms of the demarcation criterion rather than in terms of meaningfulness, since we have already seen how the principle of falsifiability furnishes a solution for the first problem. Now, unlike the positivist's, Popper's demarcation criterion cannot be construed as a criterion of meaning. He has no interest whatsoever in declaring certain theoretical systems as meaningful, and certain others as meaningless. His concern is to separate metaphysical statements from scientific ones without making the former meaningless. Moreover, Popper wants to avoid the positivist's criterion, which failed so badly to yield a good distinction between science and pseudo-science, for it ended up declaring some perfectly sensibly statements as meaningless. But more importantly, he does not think that metaphysical theories are meaningless or that they should be rejected for such reasons.[11] He thinks a theory should be weeded out from science only if it is not falsifiable. On the other hand, isolated (perhaps the vast majority of) existential statements are not falsifiable; that is, they are not testable.[12] Consequently, Popper is forced to treat these statements as metaphysical, even though he recognizes that both strict universal and strict existential statements are empirically decidable, only the former being falsifiable (i.e. empirical). In other words, Popper's criterion of demarcation violates conditions (a) and (b) above, for (1) it denies an alleged symmetry between verification and falsification and (2) it declares a falsifiable universal statement as empirical while declaring its (existential) logical equivalent as metaphysical. Since Popper's demarcation criterion is not a criterion of meaning, he can have this result without being committed to deny that universal statements can be logically transformed into equivalent existential statements.

It is worth mentioning that the criterion of demarcation applies to theoretical systems rather than to singular statements taken out of their context,[13] and that under this proviso the asymmetry between universal and existential statements is beneficial to the process of scientific theory-testing (recall, one more time, that it dissolves the problem of justifying the truth of universal statements). In section 15 of *LScD* Popper stresses what he takes as a crucial claim about the nature of this process. Since laws of nature are mostly expressed by strictly universal statements, it turns out, by way of logical equivalence, that 'they can be expressed in the form of negations of strictly existential statements or, as we may say, in the form of *non-existence statements*'.[14] This feature supports his contention that laws of nature can be better understood as 'prohibitions', and that they do not assert the existence of anything. Quite the contrary, 'they insist on the non-existence of certain things or states of affairs', which is precisely what accounts for their falsifiability.

Falsifiability 43

On the other hand, some existential statements are testable (that is, falsifiable), and hence empirical only if they are part of a larger context that is testable. Take, for example, the statement, 'There exists an element with the atomic number 72':

> It is scientific as part of a highly testable theory, and a theory *which gives indications of how to find this element.* If, on the other hand, we took this existential statement in isolation, or as part of a theory which does not give us any hint as to how and where this element can be found, then we would have to describe it as metaphysical simply because it would not be testable: even a great number of failures to detect the element could not be interpreted as a failure of the statements to pass a test, for we could never know whether the next attempt might not produce the element, and verify the theory conclusively.[15]

Then, the decidability of existential statements will depend on whether they are part of a broader context that gives us indications about how to test them, and so will determinations of their empirical status. Testability has nothing to do with the actual difficulties of falsification, for Popper will be willing to grant that the logical impossibility of verifying a universal statement is the exact analogue of the impossibility of falsifying its (existential) logical equivalent. This point is important because it is usually a source of confusion for the positivist who accuses Popper of neglecting the *logical* symmetry between universal and existential statements. Indeed, nothing could be more wrong, for Popper does not challenge such symmetry,[16] no matter how far he is willing to go in his efforts to block the typical positivist criticisms against his criterion of demarcation.

To conclude: the asymmetry between falsification and verification is not mysterious when one sees that a class of singular observation statements suffices to falsify a universal law, but it cannot ever suffice to verify it in the sense of 'establishing' it. This class, of course, can verify an existential statement but it cannot falsify it. The objections of the verificationist seem to ignore the fact that Popper treats some existential statements as metaphysical not because it is difficult to falsify them but because 'it is *logically impossible to falsify* or to test them empirically'. Moreover, there are aspects in which the problems of verification and falsification are symmetric, as illustrated above, but the existence of certain symmetries in this area does not preclude the existence of a 'fundamental' asymmetry 'any more than the existence of a far-reaching symmetry between positive and negative numbers precludes a fundamental asymmetry in the system of integers: for example, that a positive number has real square roots while a negative number has no real square root'.[17]

3.2 Falsifiability defined

As seen, falsification relies exclusively on deductive methods and procedures. If plausible, the story manages to explain how the process of selection and improvement of theories takes place, as the result of a continuous exchange between general tentative hypotheses and experience. On Popper's view we face the task of finding appropriate criteria to determine whether a theory belongs to empirical science, without having to make decisions about the semantic status of its statements. Such criteria must meet two requirements: (a) be non-inductive and (b) give us precise indications to distinguish between the typical statements of empirical science and other type of statements. The first requirement is satisfied by using an approach in which *only deductive logic* suffices to evaluate a given system of statements and perform epistemological and methodological operations, the second by formulating a principle of testability which is independent of a theory of meaning. Popper defines this principle in the following words:

> [*Fsb*1] A theory is to be called 'empirical' or 'falsifiable' if it divides the class of all possible basic statements unambiguously into the following two non-empty subclasses. First, the class of all those basic statements with which it is inconsistent (or which it rules out, or prohibits): we call this the class of *the potential falsifiers* of the theory; and secondly, the class of those basic statements which it does not contradict (or which it 'permits'). We can put this more briefly by saying: a theory is falsifiable if the class of its potential falsifiers is not empty.[18]

Let me clarify the key ideas in this formulation of falsifiability. To begin with, the notion is introduced and defined as a potentiality, a property that theoretical systems must have if they are to be considered as part of empirical science. The criterion does not impose the requirement that these systems (taken as a whole) or their component statements (taken particularly) be in fact falsified, but only that they have the feature of falsifiability; that is, that they might, in principle, contradict the possible world of experience. Such a clash, on the other hand, is neither expected nor required; it suffices with its being conceivable as a *future* possibility.

As stated in the definition, an empirical theory has to divide the class of all possible basic statements into two non-empty subclasses: the class of prohibited statements and the class of permitted statements. But how should we understand the locution 'basic statement'? For Popper the class of all basic statements includes '*all self-consistent singular statements* of a certain logical form – all conceivable singular statements of fact, as it were'. This class comprises some statements that are mutually incompatible.[19] In the alternative

Falsifiability

formulation in the last clause of *Fsb*1, he says that the system of statements (that is, the theory) is by itself a class, and it is prohibited that one of the classes related to it (the class of potential falsifiers) be a null class. In this approach to the problem, Popper is content with the non-emptiness of the class of prohibited statements. With regard to the class of compatible statements he does not make any further refinement, nor does he state any exigency (except that it cannot be empty either, if the theory is consistent), since, as he insists, theories do not make any assertion about permitted statements; they do not tell us, for example, whether those statements are true, and *a fortiori*, theories cannot give us factual information by means of this class.

A first-glance inspection of both the class of permitted statements and the class of prohibited statements yields the following results. The two classes are not complementary; that is, any increase in the size (cardinality) of one of them is not always followed by a corresponding decrease in the size (cardinality) of the other.[20] On the other hand, the class of potential falsifiers is not a proper subclass of the class of statements that comprise the theory. What I take the definition to say is that the class of potential falsifiers (call it *PF*) is a subclass of the class of all possible basic statements, and that it has certain logical relationships with respect to the system of statements that comprise the theory. For example, it may be the case that the theory (if falsifiable) entails *PF*. Furthermore, the class of potential falsifiers and the class of permitted statements are subclasses of the class of all basic statements, and so is the set of statements that comprises the theory.

To make the principle of falsifiability operative it suffices to test some of the components of a theory. Obviously, hypothetic or predictive statements that are tested are deductively related to the theory, and for this reason we only need to find, by empirical means, at least one observation statement that contradicts any of the particular consequences deduced beforehand.[21] However, in a sort of anticipation of his forthcoming amendment of *Fsb*1, Popper warns us that a single isolated observation is never enough to establish the falsity of a theory.

3.2.1 Falsifiability and falsification

In the previous section, I commented upon the definition of 'falsifiability' in terms of a disposition that determines the empirical character of theoretical systems. Falsifiability is defined as a property of any theory that belongs to empirical science, but given a general theoretical framework no empirical research is required to determine whether a system is falsifiable.[22] Falsification, on the other hand, is an epistemological act that involves methodological operations; it brings into the picture facts from the world of possible experience (i.e. test or experiment results) and therefore presupposes observation and requires an informed decision on the part who performs the evaluation, as a result of which he/she regards a theoretical system as falsified, that is, he/she accepts that it has clashed with the world of experience. In other words, falsification has occurred if we accept basic observation statements that contradict a theory. But it is necessary to introduce some rules to determine whether this is the case. To begin with, Popper explains that a singular observation statement is not sufficient to conclude that a falsification may take place (nor that it has taken place, if accepted), because 'isolated non-reproducible facts lack meaning for science'. On his view, we take a theory as falsified only if we discover a reproducible effect that refutes it. Accordingly, a low-level hypothesis that describes such an effect needs to be proposed and corroborated if we are to accept falsification.[23]

On the other hand, the falsifying hypothesis must be empirical (hence, falsifiable itself); that is, it must stand in a certain logical relation with the class of basic statements. In addition, one can see that basic statements play a twofold role. The whole system of all logical possible basic statements gives the logical characterization of the form of empirical statements; and the accepted basic statements yield the corroboration of the falsifying hypothesis, grounding our decision to declare the theory as falsified. One can say, then, that every instance of falsification is at the same time an instance of corroboration, but as explained before, each of these outcomes applies to different sets of statements. The former affects the theoretical system, the latter the deduced hypothesis (or perhaps its negation).

As Popper himself acknowledges, conclusive or definitive falsification instances if at all possible are often not free from controversy or from temporality.[24] Moreover, he has never held the view that a scientist should reject a theory on the basis of a single falsifying observation. This would make his position vulnerable to the objections of the conventionalist. Suppose that the prohibited basic statement describes a particular phenomenon that is never replicated again. Under these circumstances it seems legitimate to cast doubt upon such falsification due to its localized character. But falsification is an epistemological act (based on methodological rules adopted by us) by means of which we declare a theory as refuted because it has clashed with experience in some important (and wide enough) area, so that the exclusion of the tested portion of the system seems the only reasonable move. Besides, in order to entertain reasonable doubts we need to accord a minimum degree of universality to the falsifying statement. Finally, since science is open and public it necessitates the repeatability of the falsifying fact or experiment, by any adequately trained scientist so that he can assert its veracity and legitimacy.

3.2.2 Falsifiability: second definition

In defining *Fsb*1 Popper is not very clear with regard to what we should expect to find in the class of potential falsifiers, nor is he precise about the required level of generality of the observation statement, when this plays the role of a potential falsifier.[25] He admits that one can err when identifying candidates as potential falsifiers, and he recognizes that one can raise diverse objections about the nature, legitimacy and degree of strength of testing statements. On the other hand, as he moves on in the general discussion of falsifiability, the conclusion that a single observation statement does not suffice to refute a theoretical system quickly emerges. This shortcoming, I believe, leads Popper to amend his earlier definition and to introduce *Fsb*2, a definition of 'falsifiability' that is based on the use of two technical terms: 'occurrences' and 'events'. A more precise distinction between these terms drastically changes the qualifications required for an observational statement to count as a potential falsifier (it has to describe an occurrence) and, at the same time, expresses the criterion in a more 'realistic' language.[26] In the revised definition of 'falsifiability', Popper demands the existence of reproducible effects of a kind that contradicts a theory in order to consider it as falsifiable in either the strong or the weak sense of the term.

Let us look briefly at how to use the newly introduced terms. An occurrence is a phenomenon that takes place in a particular spatiotemporal region and can be singular, plural, brief or temporally extended. By extension, any reference to the occurrence p_k (where p_k is a singular statement and the subscripted 'k' refers to the individual names or coordinates which occur in p_k) can be successfully (and more clearly) made through the expression P_k (where P_k designates the class of all statements equivalent to p_k). According to Popper, the foregoing rule of translation gives sense to the claim that an occurrence P_k contradicts a theory t, for it brings into play all statements equivalent to p_k. Likewise, an event is 'what may be *typical or universal* about an occurrence, or what, in an occurrence, can be described with the help of universal names'.[27] An event is not a complex or a protracted occurrence, but a class of occurrences that is independent of particular individuals or space–time coordinates. For this reason events are free of the spatiotemporal limitations that hinder the appeal to basic occurrences to characterize a class of potential falsifiers. Indeed, if one wants both a minimum degree of generality in the falsifying statements and to leave some room for intersubjective testing, events have better qualifications to do the job. In addition, talking in terms of events requires reference to at least one class (instead of isolated individuals): the class comprising the occurrences that differ only in respect of the individuals involved (and their spatiotemporal determinations); that is, the class of all the statements which describe the different occurrences that instantiate the event.[28]

I find it very curious that most of the commentators simply ignore the changes in the definition of 'falsifiability' prompted by this reference to 'occurrences' and 'events', and that Popper himself rarely mentions the point again. My surprise stems from the fact that $Fsb2$ immunizes falsifiability against some of the objections of the conventionalist and most of the charges of subjectivism levelled to the earlier version of the criterion.[29] To see how, consider that an event, in so far as it represents what is typical and universal in an occurrence, can express multiple and diverse instances of such occurrence. Moreover, since homotypic events – that only differ in regard to the spatiotemporal coordinates where they take place – comprise a class by themselves (a 'homotypic' class), the total abstraction of these determinations will give rise to a class of classes. In other words, an event, even a singular one, contains one or various classes of occurrences which are described by a class of equivalent statements. Thus a good potential falsifier should prohibit not only one occurrence, but at least one event. $Fsb2$ is stated as follows:

> [$Fsb2$] We can say of a theory, provided it is falsifiable, that it rules out, or prohibits, not merely one occurrence, but always *at least one event*. Thus the class of the prohibited basic statements, i.e. of potential falsifiers of the theory, will always contain, if it is not empty, an unlimited number of basic statements; for a theory does not refer to individuals as such. We may call the singular basic statements which belong to *one* event 'homotypic', so as to point to the analogy between *equivalent* statements describing *one* occurrence, and *homotypic* statements describing one (typical) event. We can then say that every non-empty class of potential falsifiers of a theory contains at least one non-empty class of homotypic basic statements.[30]

We must not forget that theories refer to the world of possible experience exclusively by means of their class of potential falsifiers. It is easy to see that according to $Fsb2$, the class of potential falsifiers of purely existential statements, tautologies and metaphysical statements is empty. None of them prohibits a single event, and hence they assert too little about the class of possible basic statements; by contrast, self-contradictory statements assert too much: any statement whatsoever (any event) falsifies them.

Perhaps this is the right time to make more precise the idea of informative or empirical content, as a notion tied to the class of prohibited events (or potential falsifiers) and to make sense of Popper's curious definition according to which a theory gives more information about the world of possible experience in so far as it prohibits more; that is, in so far as the size of its class of potential falsifiers is larger. Let us set aside, for one moment, the problem of whether we

Falsifiability

have an accurate way of estimating the measure of a class of potential falsifiers and making precise comparisons among them. In the spirit of *Fsb2*, any theory t_2 which has a larger class of potential falsifiers (say, than t_1) has more opportunities to be refuted by experience and gives more information about the world of possible experience than t_1: 'Thus, it can be said that the amount of empirical information conveyed by a theory, or its *empirical content*, increases with its degree of falsifiability.'[31] Consequently, if the class of prohibited events of a theory becomes larger and larger, it will be relatively easy to falsify it, since it allows the world of experience only a narrow range of possibilities. In the quest for knowledge, one should aim for theories of such a high empirical content that any further increase would bring about the theory's actual falsification.

Unfortunately, there is no way of giving precise estimations of the size of a class of potential falsifiers, and this is a shortcoming of both *Fsb1* and *Fsb2*. Obviously, such a size is not a function of the number of prohibited events, let alone of the number of statements that describe these events (which, on the other hand, are infinite). For logical elementary operations (e.g. the co-obtaining of a prohibited event with any event whatsoever yields another prohibited event) would make it possible to increase artificially the cardinality of the class of events, hence the size of the class, when measured by this device. To put this another way: the mere counting of prohibited events is not an acceptable way of estimating the degree of falsifiability of a theory. In fact, Popper recommends using the subclass relation to make sense of the intuitive more or less that the comparison presupposes. But this way of comparison restricts one to consider theories that refer to the same aspect of reality, and so which are genuine rivals. These theories have to be formulated in such a way that their respective classes of potential falsifiers hold a relationship in which one of them contains as a proper subclass the other. Setting aside the issue of whether mutually incompatible but self-consistent theories can talk about the same aspect of reality in a competitive way, if the subclass relation does not obtain, the comparison simply cannot be carried out.

Whether the falsifiability criterion is applied to statements, or whether it is used to classify theories, Popper suggests a simple diagram to illustrate the different degrees of falsifiability that can be ascribed to several types of theories and rank any testable candidate according to its relative standing with regard to the extreme values. Suppose we have a horizontal continuum. We may label the extreme points of this continuum as 0 and 1. The former will indicate zero degree of falsifiability, the latter will indicate the maximum degree of falsifiability. According to *Fsb2*, it is easy to see that merely metaphysical and tautological theories have a zero degree of falsifiability, while only self-contradictory theories may have the maximum degree of falsifiability. All theories and statements that belong to empirical science are to be found in any intermediate point between 0 and 1, with the exclusion of both extremes. If one needs to compare statements or theories regarding their respective degree of falsifiability, then one would have to point to the location of those statements or theories in the continuum and attribute a higher degree of falsifiability to those which are not tautological and are closer to the extreme location indicated by the point marked with 1, without ever coinciding with it. All empirical theories have a degree of falsifiability that is greater than zero and lower than one, as is represented below:

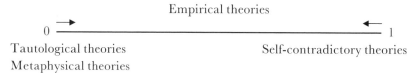

Figure 3.1

Figure 3.1 depicts correctly the comparative representation of falsifiability that Popper gives in section 33 of *LScD*. Again, the degree of falsifiability *(Fsb)* of an empirical theory *(e)* is always greater than zero and less than one: $Fsb(e) > 0$ and < 1. Based upon the relationships that can be established among different types of statements, it is possible to infer that the formula expresses, at the same time, the requisites of coherence and falsifiability, since for practical purposes it excludes self-contradictory theories. Though *prima facie*, it may look as though Popper's ideal of science would give its highest appraisal to self-contradictory theories, after careful consideration one understands that this is a mistake. Indeed, he prefers theories that are easily testable and that have the highest possible degree of empirical content. Moreover, if possible, he commends that we make any reasonable attempt to increase more and more this feature until any further change results in falsification. But this procedure has nothing to do with making the theory to be tested self-contradictory (which, incidentally does not require a great effort). To see why, notice that falsified theories do have empirical content, whereas self-contradictory theories have almost none, since they are, in general, non-informative or overly so. In other words, notice that increasing a theory's degree of falsifiability increases its empirical content, whereas making a theory self-contradictory does not.

3.3 Criticisms of Popper's notion of falsifiability

Criticisms of Popper's theory of falsifiability are abundant. They address almost every conceivable aspect of this notion, starting with its formal conditions, then dealing with alleged defects in the definition and culminating in a dismissal of the outcomes of its application. For some authors falsifiability fails to provide a suitable criterion of demarcation; for others, it suffers from the same shortcomings as does the rival and discredited notion of verifiability; still others think that the whole project of characterizing empirical science with its help is simply misconceived. Controversial as this problem is, I believe that the notion of falsifiability is quite tenable. In what follows, I will discuss the most important objections levelled against falsifiability (including the ones advanced by his former pupils Imre Lakatos and Paul Feyerabend), recast Popper's replies and develop a more charitable reading of this important concept.

3.3.1 Kuhn: testing vs puzzle-solving

T.S. Kuhn raised one of the most often-discussed objections to Popper's theory. Moreover, Kuhn's criticism (directed at the very heart of the doctrine: namely, the notion of falsification) is part of a family of objections that includes Lakatos and Feyerabend as its most conspicuous representatives. In a nutshell, these authors argue that any falsificationist methodology would be either naïve or dogmatic unless it is turned into a sophisticated methodology by appropriate modifications.[32] With minor (negligible) differences among them, they support this argument on a twofold foundation: first, they think that strict falsificationism is committed to assuming that test-statements[33] (dealing with observation or experiment) are infallible or that there cannot be falsification; second, they object that according to Popper's theory one is bound to accept these statements by convention, namely by arbitrary decisions. In sum, they see Popper's theory of falsification as lying between the Scylla of dogmatism (being forced to bestow infallibility on test-statements) and the Carybdis of scepticism (given that test-statements are accepted conventionally). Let us look at Kuhn's objection in more detail.[34]

Kuhn's general conclusion is that 'no process yet disclosed by the historical study of scientific development at all resembles the methodological stereotype of falsification by direct comparison with nature'.[35] Kuhn's argument can be summarized as follows:

1. Though anomalies are precisely counter-instances to theories, scientists never treat them as such.
2. Scientists declare a theory invalid and are ready to reject it only if an alternative candidate is available to take its place (this is the process he describes as a paradigm shift).
3. Although experiment and testing play an important role in the rejection of theories (Kuhn clarifies), such a role is not nearly as essential as the role of paradigms.

4. The actual rejection of a theory is based on something more than a comparison between the theory and the world; it involves also the comparison of competing paradigms and the assessment of the way they solve the puzzles that drive the interest of the scientific community.

What makes Kuhn's argument illuminating for some readers is that it is presented along with several points of agreement between his views and Popper's.[36] According to this list, they both are concerned with the process that leads to the acquisition of scientific knowledge instead of the logical structure of theories; both appeal to history in search of data for the facts of science; both reject the view that science progresses by mere accumulation and emphasize the revolutionary character of the processes of theory replacement. In addition, they both accept the interdependence of theory and observation; neither of them believes that a neutral observation language is possible; and they both reject empiricism.[37] The main points of disagreement are Kuhn's 'emphasis on the importance of deep commitment to tradition'[38] and his disavowal of naïve falsificationism.

For Kuhn, the testing of hypotheses – rather than being done with the purpose that Popper attributes to this process – is conducted with the aim of connecting a scientist's conjectures with the corpus of accepted knowledge; to determine whether his guesses fit the accepted trends of normal research. If the hypothesis survives, the scientist has solved a puzzle; if it does not, he still can look for an alternative path within the same paradigm. This is the kind of activity that is performed in normal science, where the tests are not directed against the current theory. It is the scientist who is challenged, not the current theory. Further, the conjectural solution to the puzzle may need to be checked in its turn, but what is tested in this case is the practitioner's ingenuity and ability to suggest a good solution and not the theory.[39] In addition to this difference, Kuhn believes that Popper is wrongly ascribing to the entire scientific enterprise (best characterized with the help of the idea of normal science) what is peculiar just to revolutionary science. Again, the testing of theories is pertinent only in extraordinary research when scientists attempt to solve a prior crisis or want to decide between competing paradigms. Because normal science discloses both the points where testing is required and the ways to perform the tests, the right criterion of demarcation should be found within it, and not in testability as Popper holds.

According to Kuhn, Popper is not a naïve falsificationist, though he may legitimately be treated as one. This type of treatment is justified since Popper's admission of fallible test-statements is 'an essential qualification which threatens the integrity of his basic position'. Moreover, Kuhn complains that, although Popper has barred conclusive disproof, 'he has provided no substitute for it, and the relation he does employ remains that of logical falsification'.⁴⁰ If test-statements are fallible, falsification cannot be conclusive, because fallible test-statements are used as premises in the falsifying arguments. This makes falsification fallible (as are its basic statements). Kuhn believes that, contrary to appearances, Popper advocates a methodology of conclusive falsifications and that there is no alternative to falsification as disproof. This is problematic for Popper since he is concerned not only with providing a criterion of demarcation, which could be achieved by a purely syntactic method, but also with the logic of knowledge that investigates the methods employed in the tests to which new ideas must be subjected. Kuhn raises another problem related to the way test-statements are accepted on the basis of actual observations and experiments. He points out, correctly, that the relation between test-statements and hypotheses is a relation between statements, and hence purely syntactic. But he thinks that syntactic (purely logical) considerations alone do not suffice for testing. In actual research, Kuhn argues, we must consider also the semantic (epistemic) relation between test-statements and observations because we do not merely want to relate sentences derived from a theory to other sentences. Moreover, Popperian falsification, if correctly construed, should function in the epistemic and methodological context. But, in Kuhn's opinion, Popper evades clarifying whether falsification can function in those contexts. Accordingly, Kuhn asks:

> Under what circumstances does the logic of knowledge require a scientist to abandon a previously accepted theory when confronted not with statements about experiments but with experiments themselves: Pending a clarification of these questions, I am not clear that what Sir Karl has given us is a logic of knowledge at all. In my conclusion I shall suggest that, though equally valuable, it is something else entirely. Rather than a logic, Sir Karl has provided an ideology; rather than methodological rules, he has supplied procedural maxims.⁴¹

In other words, Popper has not provided a Logic of Discovery, as promised. Perhaps because he rejects 'psychologism' he fails to see that Logic of Discovery and Psychology of Research, far from being in conflict, could (jointly) explain the processes of science. If Kuhn's analysis of the role of testing in science is correct, Popper's mistake is having transferred some special characteristics of the daily practice of research to the extraordinary episodes where a scientific paradigm is superseded by another. It is here where the psychology of knowledge can help us out. We may explain the growth of science by appealing to common elements induced during the practice of normal science in the 'psychological make-up of the licensed membership of a *scientific group*'.⁴² To solve this shortcoming, Kuhn proposes to replace Popper's logic of discovery and the deductive testing of theories with his psychology of research and his emphasis on puzzle-solving. Because logic and experiment alone do not suffice, scientists are not motivated by falsification; instead they want to solve puzzles[43] and are ready to embrace a theory that offers a coherent way to achieve this end while enjoying of a good deal of respect among the community of practitioners.

I will briefly analyse Popper's replies (in 'Normal science and its dangers' and *RAS*) before moving on to my own evaluation of the debate. In the paper just cited, Popper acknowledges many of the coincidences between his views and Kuhn's, but also warns us about what he considers Kuhn's misinterpretations of his work. Popper claims that he has not overlooked normal science (in favour of extraordinary research), let alone overlooked the fact that an organized structure provides the scientist's work with a problem situation. On the other hand:

> '[n]ormal' science, in Kuhn's sense, exists. It is the activity of the non-revolutionary, or more precisely, the not-too-critical professional: of the science student who accepts the ruling dogma of the day; who does not wish to challenge it; and who accepts a new revolutionary theory only if almost everybody else is ready to accept it – if it becomes fashionable by a kind of bandwagon effect. To resist a new fashion needs perhaps as much courage as was needed to bring it about.[44]

For Popper, a good scientist must reject that kind of attitude since it is totally incompatible with a critical view and promotes indoctrination and dogmatism. Moreover, Popper thinks one ought to be sorry for any scientist who has such an attitude. He also expresses a dislike for Kuhn's notion of normal science, and his disagreement with the kind of uncritical teaching that Kuhn claims takes place in the training of 'normal' scientists. Popper also argues that the existence of a dominant paradigm and Kuhn's reconstruction of a scientific discipline in terms of a sequence of predominant theories with intervening revolutionary periods of extraordinary research, may be appropriate for a science like astronomy but it hardly does justice to history of science in large part. In particular, it does not fit the continuing interaction among several theories of matter that has been taking place since antiquity.

A big difference between Popper and Kuhn is that the latter holds that the rationality of science presupposes the acceptance of a common framework. According to Kuhn, rational discussion is only possible if we agree on fundamentals and have a common language and a common set of assumptions. But Popper has criticized and rejected this view because it degenerates into relativism. It also confuses discussing puzzles within a commonly accepted framework with the false claim that a framework itself cannot be rationally criticized. This error, which Popper calls 'The Myth of the Framework', is a logical mistake that precludes those who commit it from subscribing to an objective theory of truth. Thus, rather than being restricted to a single framework, science benefits from the comparison and the discussion between competing frameworks. To summarize, Popper writes:

> Thus, the difference between Kuhn and myself goes back, fundamentally, to logic. And so does Kuhn's whole theory. To his proposal: 'Psychology rather than Logic of Discovery' we can answer: all your own arguments go back to the thesis that the scientists are logically forced to accept a framework, since no rational discussion is possible between frameworks. This is a logical thesis – even though it is mistaken.[45]

When Kuhn asserts that no historical process resembles 'the methodological stereotype of falsificationism' by direct comparison with nature, he is right in so far as he has in mind the *stereotype* of falsification by direct comparison with nature, something about which Popper's theory says nothing. But this assessment (whether appropriate or not) is against a stereotype of Popper's view and proves innocuous when rehearsed against the real view. Perhaps Kuhn would be in a better position if he made the distinction between naïve falsificationism and fallibilism. As for the latter, Popper is a convinced defender of the view that all knowledge remains fallible and conjectural, and more importantly, that there is no final justification, either for knowledge or for falsification. 'Nevertheless we learn by refutations, i.e., by the elimination of errors, by feedback. In this account there is no room at all for 'naïve falsification''.'[46]

Given the numerous points of coincidence between Popper and Kuhn, it is not easy to evaluate the outcome of their debate.[47] Since our interest here lies within the scope of Kuhn's arguments against falsifiability, we should determine whether his criticism succeeds. It seems clear that many of Kuhn's auxiliary arguments are unwarranted. For example, since the theory of falsifiability rules out explicitly conclusive disproof, it is not proper to accuse Popper of being committed to it. Nor is it charitable to characterize Popper's theory of science as an ideology of conclusive falsifications, if Kuhn means by 'ideology' that such beliefs must be kept away from criticism.

On the other hand, what Kuhn takes as the right description of the procedures of normal science (what he claims is an advantage of his own view of science over Popper's), namely that testing involves the scientist's ingenuity rather than the scientific theory, is misconceived. It betrays a verificationist attitude: the attitude of protecting a theory (the corpus of current science) from refutation and being willing to sacrifice in its place the contingent ability of the scientist. A belief in a purported majesty of Science that requires the disposal of individual talents for its own sake goes in the opposite direction of Popper's description of science as an essentially fallible enterprise.[48] The problem, however, lies not in the incompatibility between Kuhn's and Popper's views in this respect. Rather it is that the view according to which the process of testing examines precisely the objective merits of scientific theories makes better sense of the scientific enterprise than does the sociological study of science. We can dispel Kuhn's worries about the inconclusiveness of falsification by pointing out that falsification would be fallible even if we subscribed to the naïve falsificationist's stringent requirement of infallible test-statements. As is widely accepted, error may enter into the process of falsification at many different points, so banning it from the test-statements would not secure the conclusiveness of the whole process. Furthermore, it is not difficult to answer Kuhn's rhetorical question about the nature of falsification, what he posed with the intent of denying there was an alternative to falsification as conclusive disproof. I think that Kuhn runs into this difficulty because he inadvertently ignores the distinction between falsifiability and falsification and the role that each notion plays in Popper's theory of science. To produce a satisfactory answer we only need to appeal to the distinction between these two notions. It may be true that the second notion *usually* calls for conclusive disproof. But the first notion admits of more interpretations and, in particular, it gives us a way to cast doubt on the epistemic status of a theory and it promotes its rational discussion. Let us turn now to Kuhn's main argument.

To begin: premise (1) is highly questionable. Assuming that what Kuhn calls 'anomalies' represent properly a falsifying hypothesis, the claim seems either false or self-defeating. It is obviously false that scientists *never* treat counter-instances as such. But it is also self-defeating to hold that though falsifying hypotheses do refute a theory, scientists overlook it. Perhaps Kuhn may have in mind a less controversial use of (1). For example, scientists are prone to ignore anomalies until there is no alternative but to acknowledge them (either because a new theory becomes available, or because it is impossible to deny the clash). But this interpretation of (1) talks about what scientists may or may not do, instead of telling us what it would be rational to do. It is precisely rationality with which Popper's theory is concerned. In other words, Kuhn's claim in (1) may contain a true description of the behaviour of scientists who face a refuting instance, but poses no challenge to Popper's notion of falsifiability which might still be right despite the actions of researchers.

Kuhn's second premise is similarly flawed. The claim that scientists refuse to declare a theory refuted until there is an alternative candidate is exaggerated and, I suspect, false. Consider a scientist who entertains a hypothesis in a completely new field. Would the scientist simply refuse to abandon his/her hypothesis were it categorically refuted by experiment because no alternative candidate existed?

While premise (3) is less problematic, it gives little support (if any) to Kuhn's conclusion. Setting aside the difficulty of giving a precise definition of what the locution 'essential role' amounts to, should one give preference to the role of paradigms over experiment and testing when evaluating and selecting theories?

Finally, Kuhn's last premise is equally disputable. There are cases of theory rejection which do not involve any reference to paradigms or ways to solve puzzles. I find it perfectly correct to reject a theory because it fails to account properly for empirical facts, regardless of the ways it may solve puzzles.

In conclusion, Kuhn's criticism is inadequate. His general criticism might be defensible, but only in so far as it targets a stereotype which Popper need not adopt. However, even if directed to the full-blown notion of falsifiability, the fact that premise (1) is false renders his argument unacceptable.

3.3.2 *Lakatos: sophisticated falsificationism*

Lakatos criticized his teacher's philosophy of science at length. Since there are only minor changes and refinements from one paper to another, I shall draw material mainly from: 'Popper on demarcation and induction' and 'Falsification and the methodology of research programmes'. In the latter work Lakatos introduces several terminological distinctions. He speaks about 'naïve dogmatic falsificationism' (which is equivalent to Kuhn's 'naïve falsificationism') for the view that all scientific theories are fallible, whilst there are infallible test-statements. He uses the locution 'methodological falsificationism' to describe a kind of conventionalism about test-statements. Then Lakatos proposes 'sophisticated falsificationism', a position that improves over the two earlier methodologies and that forms his own theory of science, cashed out in terms of how it applies to series of theories or research programmes. In Lakatos's opinion, any philosopher who takes the point of departure from inherently infallible test-statements is a 'dogmatic falsificationist'. By contrast, a conventionalist who takes as his point of departure test-statements that are deliberately agreed upon as true is a 'naïve methodological falsificationist'.

For my purposes here, a sketch of the fundamentals of Lakatos's assessment should suffice. Lakatos is not satisfied with Popper's criterion of demarcation for, if strictly applied, it leads to the best scientific achievements being unscientific. In Lakatos's opinion, psychoanalysis and Newtonian science are on a par as far as their respective inability to set forth experimental conditions that, if obtained, would force supporters of one or the other to give up their most basic assumptions.[49] The claim that any fundamental theory is conclusively falsifiable is false. Hence Popper's own philosophy is refuted. As a consequence, Popper's recipe for the growth of science (boldness in conjectures and austerity in refutations) does not yield good criteria for intellectual honesty.[50] On the other hand, Lakatos maintains that Popper conflates two different positions about falsification: 'naïve methodological falsificationism' and 'sophisticated falsificationism'. To explain these positions, albeit briefly, and to evaluate Lakatos's accusation, we first need to understand 'dogmatic falsificationism'.

'Dogmatic falsificationism' admits that all scientific theories are fallible, without qualification, but attributes infallibility to the empirical tests. It recognizes the conjectural character of all scientific theories and the impossibility of proving any theory. Lakatos thinks that such a view is justificationist, given its strict empiricism. By settling for falsifiability, dogmatic falsificationists have a modest standard of scientific honesty. They also hold that the repeated overthrow of theories with the help of hard facts produces the growth of science. According to Lakatos, two false assumptions, conjoined with an insufficient criterion of demarcation, support this position. The assumptions are:

(a) the existence of a natural, psychological, borderline between theoretical and observational propositions; and
(b) that any proposition that satisfies the (psychological) criterion for being observational (factual) is true.

Lakatos appeals to Galileo's observations of the moon and the sun to show that psychology renders false the first assumption while logic speaks against the second. His argument runs as follows: Galileo considered the old theory that celestial bodies were faultless crystal spheres refuted because he observed mountains in the moon and spots in the sun. But his observations were not observational in the sense of having being made with the pure, unaided senses. They were obtained with an instrument whose reliability was in question. Since Galileo did not have an articulated optical theory to validate his telescopic observations, nor did he have any means to confer legitimacy to the phenomena observed and to assure the reliability of his optical data, he was not in possession of an infallible test-statement. Hence, we do not have a case of refutation. Rather, we have a confrontation of Aristotelian observations (made in the light of a well-articulated theory of the heavens) against Galilean observations (made in the light of his precarious optical theory). We have two inconsistent theories. The upshot is that 'there are and can be no sensations unimpregnated by expectations and *therefore there is no natural (i.e., psychological) demarcation between observational and theoretical propositions*'.[51]

The second assumption of dogmatic falsificationism is false because there is no way to prove any factual proposition from an experiment. 'Observational' propositions are not indubitably decidable. Thus, all factual propositions are fallible. On the other hand, since propositions cannot be derived from facts but only from other propositions (a point of elementary logic that Lakatos thinks few people understand), it turns out that a clash between a theory and a factual proposition is mere inconsistency rather than falsification. In conclusion, the distinction between theoretical propositions and empirical propositions falls apart: all propositions of science are both theoretical and fallible.

Lakatos then argues that even if his evaluation of assumptions (a) and (b) above were wrong, no theory would ever meet the criterion of falsifiability in the sense of prohibiting an observable state of affairs. Lakatos tells how a Newtonian scientist confronted with the anomalous behaviour of a planet is able to engage in an endless process of inventing ingenious conjectures instead of considering every deviation from the predicted values as an instance of a state of affairs prohibited by the theory. In other words, the Newtonian scientist saves the theory from falsification by tirelessly introducing auxiliary hypotheses as needed.

In contradistinction to Popper, Lakatos thinks that irrefutability, in the sense of tenacity of a theory against empirical evidence, should become the hallmark of science.[52] Moreover, anyone who accepts dogmatic falsificationism (i.e. the demarcation criterion and the idea that facts can prove factual statements) should reach the exact opposite conclusions that Popper endorses: the dogmatic falsificationist would have to consider the most important scientific theories ever proposed as metaphysical; he would be forced to redefine completely the notion of scientific progress (denying that we have reached any real progress so far) and lastly would be compelled to admit that most of the work done in the history of science is irrational. Dogmatic falsificationism leads to rampant scepticism:

> If *all* scientific statements are fallible theories, one can criticize them only for inconsistency. But then, in what sense, if any, is science empirical? If scientific theories are neither provable, nor probabilifiable, nor disprovable, then the sceptics seem to be finally right: science is no more than vain speculation and there is no such thing as progress in scientific knowledge.[53]

A methodological falsificationist (like Popper) applies the (fallible) theories, in the light of which he interprets the facts, as unproblematic background knowledge while the testing is performed. In this respect, he uses successful theories as extensions of the senses and widens the range of theories that can be used in testing, compared with the limited range of the dogmatic falsificationist.[54] The methodological falsificationist adopts an heroic stance: in order to escape scepticism he is willing to grant observational status to a theory by accepting certain test-statements as true but fallible, without making any concession to justificationism. In this way, he appeals to the empirical basis but is not committed to its foundational character. Furthermore, he provides methodological rules that lead to the rejection of a falsified theory, but unlike the dogmatic falsificationist (who conflated rejection and disproof), he maintains that fallibilism does not put any limits on the critical attitude. Some statements are the basis of tests in spite of being fallible. While we can appreciate the courage of such a hero, from an intellectual point of view the situation is far from satisfying. After pointing out that the methodological falsificationist has to deal also with the problems of the *ceteris paribus* clause, Lakatos asks if those who advocate decisions are not 'bound to be arbitrary'.[55]

Though the Popperian thread may lead to the right methodology, both dogmatic falsificationism and methodological falsificationism are mistaken. But there is no need to despair: Lakatos's own sophisticated falsificationism (in the spirit of that thread) is supposed to solve the difficulties. Instead of focusing on isolated theories, Lakatos considers progress and testing through series of theories (what constitutes a research programme). In addition, sophisticated falsificationism reduces the conventional element in falsification rescuing methodology and scientific progress. On the other hand, falsification can be carried out only in the light of series of theories, where each subsequent element of the series is the result of adding auxiliary clauses to the previous theory or is a semantic reinterpretation of it. Only if we have a (better) competitor can we consider a theory as refuted. Otherwise, we should dismiss the clash of the theory with the experimental results since no theory is falsifiable$_2$ by any finite set of factual observations. Just as Duhem argued, there are no crucial experiments that can conclusively falsify a theory, and thus it is always possible to rescue it by introducing auxiliary hypotheses or reinterpreting its terms.[56] Finally, sophisticated falsificationism applied to test-situations that involve a series of theories is more congenial to history of science and can accommodate the 'infantile diseases' of science such as inconsistent foundations and the occasional introduction of *ad hoc* hypotheses which Popper's methodological falsificationism rules out.[57]

I shall argue that Lakatos's objections against falsifiability are unconvincing and misinterpret Popper. Let me begin with the dual contention that there are no genuinely falsifying instances and that Popperians either take test-statements as infallible (which seems to contradict the principles of rational criticism) or declare them true by arbitrary convention (which commits them to scepticism). As explained above, Lakatos argues for the first part of this contention by appealing to Galileo's astronomical discoveries which he uses both to discredit the theoretical/observational distinction and as an example of the uncritical use of a theory as background knowledge by the methodological falsificationist. Lakatos's analysis, however, is flawed in the following respects. First, the statement that Galileo was trying to refute the old theory about the perfect character of the celestial bodies is inaccurate. This issue was not central in the debate on the Copernican system which Galileo wanted to support. It is true that the old Aristotelian idea of the perfect nature of celestial bodies was in disrepute (mainly after speculations by Nicholas of Cusa and Giordano Bruno) and telescopic observations could have given basis to an empirical refutation of such an idea, but this is by no means the important part of Galileo's own report of his astronomic observations. No argument on this matter appears in the *Sidereus Nuncius*. The topic is just an aside in the *Letters on Sunspots*[58] and occupies a secondary place in some of Galileo's personal correspondence. Second, Lakatos's misgivings about the observational status of the telescopic discoveries are unjustified. He draws a sharp contrast between test-statements compatible with Aristotle's cosmology and those invoked by Galileo. The former statements, Lakatos insists, though part of a fully developed theory were obtained by 'natural' means. In contrast, the latter did not have a suitable theoretical support and were not purely 'observational'. Therefore, Galileo's telescopic observations were unreliable and theory-dependent.

But even if this reconstruction were correct, Lakatos misses two points: First, the imperfections on the surface of the moon and the sun are observable with the unaided eye and were known since ancient times. That explanations for these phenomena were false, inadequate or feeble is an entirely different problem. Galileo mentions in his second letter on sunspots that some spots had been observed as early as the time of Charlemagne, and he suggests to his correspondent an easy way to examine them with the natural senses. He reports that medieval astronomers believed that the spot reported in the *Annals of French History* (Pithoeus, 1588) was produced by the interpolation of Mercury (then in conjunction with the sun) and he notices that the Aristotelian theory might have deceived astronomers into that false explanation. But he also adds that if more diligent work had been put in the task, the right explanation would have emerged. Second, although the nature of the instrument that Galileo used to make his astronomic observations combined with the nonaccessibility of the heavenly bodies might have prompted rational doubts on the reliability of the observations and their corresponding epistemological status, Galileo's analogy argument provided some rational support for the legitimacy of the observations made with such an instrument. It was evident that the telescope, when applied to terrestrial objects, increased their size by as much as a thousand times and made them appear 'more than thirty times nearer than if viewed by the natural powers of sight alone'.[59] Since these effects were related to the size and distance of the objects observed, it was sensible to think, as Galileo's argument suggests, that extremely distant objects which were visible to the unaided eye, should not increase their apparent size dramatically and, by the same token, that very distant objects – rendered visible by the instrument – were real. Only sheer prejudice about the unique nature of celestial bodies persuades Galileo's opponents of the unreliability and lack of scientific value of the telescopic observations.

Assuming my points above are well taken, they illustrate how certain test-statements do refute a theory, although its refutation may not have been the primary intention of a scientist. Setting aside the complex debate about the reliability and significance of Galileo's astronomical discoveries (as well as the problems pertaining to the Copernican theory), if the whole episode does not constitute an instance of theory refutation, then I do not know what does. On the other hand, it seems odd to claim that Galileo regarded those observations as infallible, let alone true by convention. Nor should a Popperian be committed to interpret them in those terms. Based upon his own experience, Galileo positively thought that the telescopic observations could be improved on as better-designed instruments became available, therefore there are other interpretative choices open to scientists. Finally, this historical episode can hardly give support to Lakatos's claim, according to which there is no genuine distinction between theoretical and observational propositions.[60] Thus Galileo's case will validate neither of Lakatos's rejections of the two first varieties of falsificationism. Though one can say in Lakatos's defence that some of his strictures are not mistaken. I turn now to an examination of the subsidiary arguments expounded in the first part of this section.

It is important to mention that Lakatos neglects the distinction between occurrence and event, and that this confusion affects his argument against falsification based on the alleged failure of scientific theories to forbid possible states of affairs, except in the narrow sense of theories with *ceteris paribus* clauses. Although he uses the locution 'singular event', the context of his discussion reveals that he is talking about occurrences. For these, and only these, are indexed to particular spatiotemporal regions in Popper's theory, and, as I explained in section 3.2.2 above, falsifiability needs to be reformulated in terms of events which are more general (universal) than mere occurrences. In other words, the notion of falsifiability that Lakatos is attacking under the label of 'dogmatic falsificationism' is one that Popper does not hold, except for the purposes of introducing some theoretical distinctions. On the other hand, even if Lakatos's criticism could be reworked in such a way that it applies to the right construal of falsifiability, I think it misses the point. The *ceteris paribus* clause seems to be rather a metaphysical assumption (just as the claim of the regularity of nature). The fact that a theory can be reformulated by replacing these clauses is not a problem since falsifiability, in any case, is subject to revision under Popper's view. However, if what Lakatos has in mind is that it is always possible to save a theory from falsification by introducing *ad hoc* hypotheses and making conventionalist moves (as one might think from his example of an observed black swan as unable to refute the statement 'all swans are white'), then he is simply refusing to accept the principles of Popper's theory where both these ways of defending a theory have been ruled out. As for the complaint that no falsification is absolutely conclusive, Popper recognized this point in several places and it does not require any additional comment.

Falsifiability

Lakatos's remedy for falsificationism can be disputed. To begin with, his own sophisticated methodological falsificationism is not free of conventionalist tendencies. I do not intend to offer a *tu quoque* argument against Lakatos, but since he stresses this alleged failure in Popper he can be criticized for making a similar mistake. According to Lakatos, test-statements should be accepted in accordance with a research programme. But is not a research programme accepted by agreement in the first place? Now, Lakatos admits that regarding the empirical basis, sophisticated falsificationism shares the difficulties faced by naïve falsificationism.[61] He thinks he is better off though because the conventional element in his methodology can be mitigated by an appeal to a procedure. But I fail to see how distinguishing degrees of conventionalism is going to help his case or how the 'appeal' procedure is radically different from Popper's pervasive fallibilism. If my reading of Lakatos is correct, then his own solution to the problem of the acceptance of basic statements turns out to be unsatisfactory.

Furthermore, Lakatos misinterprets Popper's stance on the reasons for accepting basic statements. Popper has not claimed that we should consider these statements as infallible by agreement. He has claimed something entirely different. He holds that test-statements should be falsifiable at their turn. But for the purposes of avoiding an infinite regress (the regress that afflicts the justificationist), we should agree (in the cases where this is a reasonable move) that some test-statements are true (an agreement that is not immune to revision).

Lakatos's brand of falsificationism is supposed to improve on Popper's theory both in its rules of acceptance and its rules of rejection. Moreover, only a theory that has more (corroborated) empirical content than its predecessor qualifies as scientific, and only if a fit competitor is proposed can a theory be falsified.[62] But the first definition is hardly a novelty in the processes of theory-selection and theory-replacement on Popperian principles, and the second is inadequate. If a theory is proven false and no alternative is available, it may be wise to retain it (as a known false approximation), but we certainly need to declare it falsified regardless of how beneficial it would be for the purposes of the growth of science, in order to have a suitable competitor. A similar problem affects Lakatos's charge that Popper is forced to eliminate any theory that has been falsified. This overstates Popper's view. Such a theory need not be eliminated in any situation in which it would be more sensible to retain it. But this should not affect its logical status: though retained, it is still a *refuted* theory.[63]

Although this is a minor point, Kuhn's and Lakatos's independent attempts to falsify Popper's theory of science are misconceived and misleading. Popper's theory is not intended as a scientific theory, but rather as a metaphysical or a methodological theory about science.

Finally, many of Lakatos's points are supported by textual evidence taken from Popper's work, but the quotations ignore the relevant background and are taken in isolation of clarificatory developments. Consider a simple example given by the motto: science progresses by means of conjectures and refutations. It is frequently used and given as a short statement of Popper's views. But is misleading to take it out of context, ignoring all the qualifications that the phrase has prompted, in order to show that as a 'pattern of trial-by-hypothesis followed by error-shown-by-experiment' breaks down.

3.3.3 Feyerabend: epistemological anarchism

Feyerabend's criticisms of falsifiability follow a similar pattern to the objections raised by Lakatos to whose revision of Popperian ideas he is sympathetic.[64] However, Feyerabend's own epistemology is closer to open irrationalism or at least to a sort of liberalism as is expressed by his motto 'anything goes'. For Feyerabend the progress of science is fostered by epistemological anarchism, a view that advises the scientist to take advantage of any methods that seem suitable for the particular conditions he is facing (sometimes labelled also as methodological opportunism). It might lead, in some cases, to reject theories, and in other cases, to defend them. This anarchism, though, has little in common with political anarchism. It only resembles the latter in so far as it eventually recommends going against the status quo and refusing to abide by any set of rules. According to Feyerabend, any fixed set of methodological rules hinders the progress of science and could even prevent its very birth.[65] That is precisely the problem of naïve falsificationism, the essential part of Popper's methodology on this reading. Feyerabend reacts against the idea of conclusive falsifications and the corresponding strict principle of falsification, maintaining that this principle 'is not in agreement with scientific practice and would have destroyed science as we know it'.[66]

According to Feyerabend, the demands of falsification (judge a theory by experience and reject it if it contradicts accepted basic statements) are useless because no scientific theory can agree completely with the known facts. In addition, if Hume was right in showing that no theory can be derived from experience, we better drop from our epistemology any requirement about the dependence of theories upon facts along with dictates about confirmation and empirical support (if made under the assumption that theories can agree with facts). We should also revise our methodology, making room, *inter alia*, for unsupported hypotheses. Accordingly: '[t]he right method must not contain any rules that makes us choose between theories on the basis of falsification. Rather, its rules must enable us to choose between theories which we have already tested *and which are falsified*.'[67]

Falsifiability

Feyerabend also holds that there is no clear-cut distinction between theories and facts, nor is there a univocal definition of 'observations' or 'experimental results'. Moreover, all the material with which a scientist works (including laws and mathematical techniques) is indeterminate, ambiguous and parasitic upon the historical context. This material is contaminated with (theoretical) principles (often times unknown), and extremely difficult to test if known. For example, test-statements (in so far as they are 'observational') are the result of the combination of 'sensation' and 'interpretation'.[68] But the interpretation of sensations is made in accordance with some theory. This interpretation is not the result of a conscious act since one is often not aware of distinguishing between 'sensation' and 'interpretation'. Feyerabend stresses that noticing a phenomenon and expressing it with the help of the appropriate statement are not two separate acts, but one. By contrast, producing an observation statement consists of two very different psychological events: (1) a clear and unambiguous sensation and (2) a clear and unambiguous connection between this sensation and some parts of the language. Inspired by Bacon, Feyerabend calls 'natural interpretation' those mental operations (so closely related with the senses that they come just after their operation) and he explains that many epistemologists have considered them as prejudices (or *a-priori* presuppositions of science) that need to be removed before any serious examination can begin. Nonetheless, what usually occurs is that they are replaced by different (sometimes competing) natural interpretations.

Natural interpretations are a constitutive part of observational fields and thus part of science. Without them, Feyerabend maintains, one could neither think nor perceive. Consider what would happen to anyone who attempted to approach the world in such a way: 'it should be clear that a person who faces a perceptual field without a single natural interpretation at his disposal would be *completely disoriented*; he could not even *start* the business of science'.[69] For this reason, he concludes, the intention to start from scratch is self-defeating. On the other hand, it is not possible to unravel the cluster of natural interpretations without circularity, unless one appeals to an external measure of comparison. *A fortiori*, it is not possible to criticize test-statements (which are firmly entrenched in natural interpretations) without introducing a new interpretative theory. But then, conflict between a test-statement and a theory need not lead to conclusive falsification (or to the elimination of the natural interpretation that causes trouble) but may lead to replacement of the offensive interpretation for another just 'to see what happens'. Hence, Popper's criterion of demarcation – with its dependence on test-statements and its reliance on critical discussion – is inapplicable to science. Let us see why.

According to Feyerabend, the main principles of Popper's theory of science 'take falsification seriously; increase content; avoid *ad hoc* hypotheses; "be honest" ... and, *a fortiori*, the principles of logical empiricism ... give an inadequate account of the past development of science and are liable to hinder science in the future'.[70] Both principles fail to account for the development of science because science is more 'sloppy' and 'irrational' than any methodological reconstruction can capture; they hinder the future progress of science because attempts to make science more precise and rational wipe it out. Given my concern in this chapter, let us isolate Feyerabend's qualms about falsifiability. Briefly stated, he claims that it is not possible to exercise critical rationalism on test-statements because they are theory-dependent. If this is the case, the Popperian would have to embrace naïve falsificationism which would give us a very impoverished methodology and would leave us with no conclusive falsification. As an alternative to this view, one must use 'irrational' and 'anarchistic' methods such as propaganda, rhetoric and psychological tricks. Feyerabend thinks (and so does Kuhn, as we saw) that a method that relies on falsificationism or sheer criticism is impossible due to the theoretical frameworks that are used to interpret sensations and to arrive at test-statements. Thus, the best alternative to naïve falsificationism is epistemological anarchism.

Feyerabend does not offer a single structured argument to sustain the aforementioned criticisms. He claims that history of science provides extensive support for them. Accordingly, he proceeds to examine several historical cases that allegedly illustrate how scientists frequently replace natural interpretations for other competitors to fix the theory; how they introduce hypotheses *ad hoc*, etc. instead of behaving like falsificationists who declare the bankruptcy of the theory when faced with the first difficulty. For example, when Galileo dissolves the traditional observational objections against the motion of the earth (which should qualify as test-statements, at least under Aristotelian physics) he admits the correctness of the sensory content of the observation made and questions only the reality of that content. He claims one must try to uncover its fallacy, that is to find a way to make the observation fit. To solve the difficulties, Galileo replaces the natural interpretation that causes trouble to the Copernican theory for a different interpretation; in so doing, he introduces a new observation language.[71]

Since Feyerabend attributes such importance to historical cases in support of his criticisms of falsification, I must comment on the tenability of such analysis. Feyerabend's favourite example is Galileo, whom he sees as an 'opportunist' (a label that if rightly attributed to the Italian scientist will discredit Popper's reference to Galileo's observations of the heavens with the telescope as an instance of a case in which a refutation led to a revolutionary reconstruction).[72] On Feyerabend's analysis, Galileo's defence of the Copernican system of the world is based on propaganda, methodological opportunism and epistemological anarchism. Feyerabend's argument can be reconstructed as follows. The Copernican system was able to explain and predict the positions of the planets rather well. But the changes in the apparent size and brightness of the planets, which were accessible to naked-eye observation, did not conform to the values predicted by the Copernican theory. Therefore, there were test-statements (relative to the size and brightness of the planets) that contradicted the Copernican theory. Since there is no observational error involved (for naked-eye observations could have been repeated *ad nauseam* with the same values that refute Copernicus) the only conclusion left for the naïve falsificationist is that Copernicus's theory is false and must be rejected.

Feyerabend's story explains why Galileo thought that the Copernican system was false and rejected it. Then, after his observations with the telescope, Galileo noticed that the changes in the size and brightness of the planets were, indeed, in accordance with the Copernican theory. Since Galileo considered the telescope a 'superior and better sense', he changed his mind based on the new observations and ceased to consider the Copernican theory as 'surely false'. Furthermore, Galileo solved the tension between the reliability of naked-eye observation (which was out of question, from the point of view of the tradition) and the reliability of the observations made with his recently developed tool by means of a trick. Galileo limited himself to saying (dogmatically) that the telescope was reliable, disregarding the background knowledge and the common sense prevailing in his epoch. Thus we have a case of falsification based on naked-eye observation that was not regarded as conclusive and dismissed with no further ado until independent and new test observations (that supported its dismissal) became available. But the example also shows that every falsification is conditional on the truth of the test-statements, and that if we accept the test-statements as fallible, then the falsification performed on their basis should also be deemed fallible.

There are two components in Feyerabend's argument. The first is directed against Popperian epistemology and (simply stated) says that Popper (as a naïve falsificationist) is committed to the view that test-statements are infallible. The second concerns Feyerabend's interpretation of Galileo's arguments to support the Copernican system and Galileo's own appraisal of the telescopic observations. I will show why both components are ill-conceived. Let us start with the accusation of Popper's commitment to the infallible character of test-statements. It cannot be supported on textual evidence, nor does it follow from Popper's theory. I grant that, *prima facie*, falsification seems to require the stability of the truth value of test-statements and that some readers find perplexing any view that combines *fallible* test-statements – which truth values are subject to revision – with falsification. But I think there is nothing to be perplexed about. As with any human enterprise, falsification is also liable to err at any step. According to Popper, under normal circumstances there are clear instances of falsification and they are not incompatible with temporal agreement about the truth-value of test-statements nor with the occasional necessity of revising our decision about such appraisals. On the contrary, this procedure is an advantage of critical rationalism. In sum, the fact that fallible test-statements retransmit fallibility upon the corresponding falsification poses no problem for Popper's theory, since he has stressed both that test-statements should be amenable to empirical test and that one should be able to recheck the procedures in the testing process without having ever claimed that test-statements need be treated as infallible. Furthermore, when Popper commends the testing of test-statements he does it under the assumption that it enables us to avoid observational error. An astronomer can repeat his telescope-aided observations of the heavens (as Galileo actually did) and then frame some low-level hypothesis about the observations. Popper has emphasized that science does not accept isolated test-statements. Tests are accepted together with low-level test hypotheses, which if falsified (or corroborated) will falsify (or corroborate) the upper-level hypothesis that is being tested.

Feyerabend's historical analysis of Galileo's defence of the Copernican system is found wanting. Though I am not a historian of science, studying some of Galileo's main works gives us evidence to reject Feyerabend's key claims on the following counts.[73] First, Feyerabend's statement about Galileo's disbelief in the Copernican system *prior* to his observations with the telescope is, to say the least, dubious, unless Feyerabend is referring to a very early stage in the development of Galileo's scientific opinions, in which case the statement is not relevant to the issue. In one of Galileo's letters to Kepler (dated 4 August 1597) the Italian scientist already expressed belief in the truth of the Copernican system. Moreover, Galileo made his observations with the *perspicillum* between 7 January and 2 March 1610. When he published his *Sidereus Nuncius*,[74] to report what he took as 'great and marvellous' discoveries to the academic world, he was persuaded that the Copernican system was true. In addition, he dared to make his first public statement supporting Copernicanism because he though that Jupiter's moons and the phases of Venus were empirical facts which corroborated Copernicus's theory, the latter being a clear refutation of Ptolemy's theory. While I must set aside technical details about the differences between the two competing theories (the Ptolemaic and the Copernican systems of the world), I will summarize the issue in five points.

First, Galileo was opposing Aristotelian cosmology and Ptolemaic astronomy that kept physics and astronomy as separate fields. To refute a substantial part of Aristotelian cosmology, Galileo needed observations capable of falsifying the view of celestial bodies as perfect entities. He found such tests in his observations of the moon. Second, views concerning the perfect character of heavenly bodies had nothing to do with the truth of Copernicus's system, since he remained silent in this respect. If Galileo was hesitant to subscribe publicly to the Copernican theory, it should probably be attributed to the

Falsifiability 69

lack of good empirical arguments supporting Copernicanism at that time.[75] It was not reasonable for Galileo to engage in a public dispute without good arguments. Third, I have no idea how Galileo's observations with the telescope yield any values that show how the differences in apparent size and brightness of the planets can be accounted for by Copernicus, since there is no place in the *Sidereus Nuncius* in which Galileo makes such a statement or gives any support to Feyerabend's claim. Moreover, other than Galileo's trigonometric estimations of the height of the moon's mountains and the detailed description of the positions of the moons of Jupiter (for a period of barely two months), there are no measurements (nor estimates) of the position of any heavenly body in Galileo's book. Nor are there such measurements in the rest of the Galilean astronomical treatises. Fourth, Feyerabend is surely right in his contention about the appraisal that Galileo's contemporaries could have made of the reliability of the observations (of celestial bodies) aided only with the telescope, compared to the reliability of naked-eye observations. Indeed, one could add on Feyerabend's behalf that Galileo lacked an optical theory (at least a correct and well-articulated theory) to explain the functioning of the telescope. He simply disregarded Kepler's optical theory (mentioned to Galileo by the Imperial Mathematician himself in *A Conversation with the Sidereal Messenger*, written as a follow-up to Galileo's book). But this does not show that Galileo's commendation of the telescope and his confident appeal to his findings with this scientific instrument are merely the product of propaganda, of an arbitrary decision and of 'epistemological anarchism', as Feyerabend holds. Disregarding the novelty of the scientific tool (and surely it made its results suspicious at the beginning), there were independent and well-trained scientists who confirmed Galileo's observations.[76] Fifth, though some of his contemporaries (e.g. Libri) not only discredited Galileo's observations with the telescope but also refused to use the instrument themselves (Galileo then initiated a vigorous campaign to convince his friends and neutralize the criticisms of his enemies), this fact does not show that he was relying exclusively on rhetoric or propaganda. It shows only that when there are problematic auxiliary hypotheses involved they are usually discussed in non-rational or non-orthodox ways.

To finish my discussion of Feyerabend's criticisms, I would like to add a few words on his contention that Galileo did not act as a naïve falsificationist in regard to the clash between the Ptolemaic model and his observations of the planets with the telescope. If by this claim Feyerabend means that Galileo did not consider the Ptolemaic model falsified as soon as he completed his first observation, then Feyerabend is right. (Incidentally, nor would any Popperian make such a claim.) But what happened in Galileo's case was, roughly speaking, that he identified several test-statements which conflicted with many claims of the traditional system of the world (as described above) and was able to offer better arguments to show that the theory (which was under heavy criticism from several quarters) was evidently false. It would be erroneous, then, to deny that after completing his observations of so diverse and meaningful celestial phenomena, Galileo was in a position to regard the traditional view as falsified.

Feyerabend's reference to the debate concerning the reliability of the observations made through the telescope, although historically accurate, misconstrues the facts. From such debate one might infer exactly the opposite lesson that Feyerabend draws. That Galileo's contemporaries were reluctant to accept his discoveries as the result of a legitimate method and an appropriate scientific instrument only shows that they participated in a critical discussion of test-statements. But this is what Popper's method commends. So, the whole episode can be understood (even better) from the point of view of falsifiability.

3.3.4 Derksen: falsifiability is fake cement

I will spend some time reviewing Derksen's objections and responding to them because he is an author who, in general, understands very well Popper's philosophy of science. Indeed, his argumentative strategy is to introduce the basic notions, show their interrelation, make Popper's position as strong as possible and then attack it with a charge which, if true, would be devastating: falsifiability is a fake cement because Popper is guilty of ambiguity when using this key notion in several passages. Though Popper himself has called to the attention of his readers that a correct interpretation of his notion of falsifiability must leave room for two senses (one logical, the other methodological), the charge seems, *prima facie*, correct.

Derksen considers that the essential concepts in Popper's theory of science grow organically out of the notion of falsifiability. Popper's system, Derksen continues, exhibits unity in the form of a 'great chain of concepts' sprouted to falsifiability in which the methodological coherence stands in an impressive way. The desiderata for a theory to be the most falsifiable, the best-testable, the most informative and the best-corroborated can be fulfilled at the same time. But the appearance of methodological unity dissolves and the great chain breaks, when we discover a hidden ambiguity in the concept of falsifiability:

> For instance, the best testable theory, i.e., the theory which runs the greatest risk of being refuted and so offers us the best chance to learn something, and the best corroborated theory, i.e., the theory which we have reason to believe is the closest to the truth, need not be the same theory anymore. Additional requirements might seem to restore the unity; unfortunately, they only do so when some inductive assumptions are made. That is, I

Falsifiability

argue that the impression of unity made by Popper's philosophy is due partly to a hidden ambiguity in his concept of falsifiability and partly to suppressed inductive assumptions.[77]

Having announced his critical point in this way, Derksen moves on to the expository part of his paper. He summarizes Popper's theory of science with the help of three claims and their respective elaborations. The final product looks like this:

Claim I: only from our mistakes can we learn.
Claim II: the more falsifiable a theory, the better the chance of scientific growth it offers.
Claim III: corroboration gives us a reason for believing that science has come closer to the truth.[78]

On Derksen's reading, Claim I implies that only through empirical falsifications can science grow. It also means that the only method to achieve truth in science is the method of bold conjecture followed by empirical refutation, through the critical examination of our guess. In this process, falsifiability (in the form of criticism) affords us a way of escaping from scepticism and irrationalism. Claim II means that a bolder theory can be more easily refuted but offers the greater chances of learning something new, and hence the greater chance for the growth of science. At the same time, it gives support to the assertion that the best-testable and most informative theory has also the greatest explanatory power and the greater simplicity. According to Derksen, the foregoing seems more impressive because it yields a picture of a science that grows in spite of the invalidity of induction, with the extra bonus of an attractive theory of rationality. Before moving on to the next claim, however, Derksen explains why Popper needs to amend slightly Claim I. In the revised form, this becomes: 'only from falsification, occasionally preceded by corroboration, can we learn.' The motivation for such amendment is that we need a reason to believe that we are producing better theories in the sense of making progress towards the truth. For otherwise, if we were confronted with an endless sequence of refuted theories this might mean that we have produced *ad hoc* theories, which although having an increasing degree of testability, would not make any real progress towards the truth. In short, without corroboration, as Popper writes, 'science would lose its empirical character'.[79]

Derksen contends that giving corroboration such a role begs the question, since it relies on Claim I (as well as on its amendment), and these claims assume that corroboration achieves something. A similar argument, he continues, can be run in relation to Claim II, since science would not grow in the Popperian sense, unless some attempted falsifications fail, or in other words, 'unless we occasionally corroborate a theory'.[80] In addition, Derksen maintains that Popper cannot substantiate Claim III. It is by no means necessary that the best-corroborated theory be the most falsifiable one, in the sense that it is the one which could undergo the most severe tests. It could be the case that all the risky predictions had already been tried out to no avail and that only the less risky predictions remained to be tested; that is, those that had a smaller chance of being disproved. If this is the case, falsifiability is bound to decrease after each failed attempt to falsify the theory (i.e. after every test whose outcome is corroboration). Derksen finds these results problematic. He thinks that if only the less risky predictions remain to be tested, this would prevent us from conferring upon the same theory the desirable qualities of falsifiable and well-corroborated, and he considers odd the idea of decreasing falsifiability (in the sense of testability). But we can dissolve Derksen's worries. To block the former it suffices to recall that what Derksen considers so problematic is exactly what accounts for a theory having being highly falsifiable and becoming well-corroborated (these two properties are necessary counterparts). Though it is hard to imagine the ideal case Derksen dreams of (high falsifiability *cum* corroboration), it is not impossible. However, for the purposes of science one can welcome a theory well-corroborated although its degree of falsifiability does not look too impressive, provided it is falsifiable$_1$. Nor is the problem of decreasing falsifiability very critical. As everyone familiar with Popper's philosophy knows, the idea is right and has been discussed at length in several places. In fact, Popper writes in *CR*: 'the empirical character of a very successful theory always grows stale, after a time' and makes clear that this poses no problem for his theory of science so long as we understand falsifiability as a property (in the sense of falsifiability$_1$) and not as a measure (according to the two senses carefully distinguished by him). In the former case, falsifiability$_1$ may diminish but cannot possibly become zero unless the theory under consideration becomes tautological, which does not seem to be the case for any known empirical theory.

This way of resolving the problems does not satisfy Derksen who is now ready to make his main point:

> What Popper does not seem to realize, however, is that in the meantime a number of *different* concepts of falsifiability have shown up: (a) falsifiability as informative content (i.e., as the class of potential falsifiers) and (b) falsifiability as testability (i.e., as the class of potential falsifiers which have not been tested). Expressed in these terms, we have just observed that a theory may be (c) highly falsifiable in the sense of having a high informative content, and yet (d) may not be very falsifiable in the sense that very few serious tests for the theory are left ... So, rather than having one Great Chain of concepts all connected with the one and only concept of

falsifiability, we now have (e) two different concepts of falsifiability, viz. informative content and testability, and a concept of corroboration which is complementary to testability.[81]

But there is more to be said. Even a third concept of falsifiability can be identified in Popper's philosophy: namely, falsifiability as the ease of falsification. According to Derksen the conclusion suggests itself: 'the choice of the best corroborated theory need neither coincide with the choice of the best testable theory nor with the choice of the most informative theory.'[82] At this point falsifiability melts and the Popperian finds himself with no reason to choose among the most testable theory or the best-corroborated theory, or even the one that is easier to test. The last category is important because one can imagine a case in which a bold theory might require tests so risky or so difficult to perform that one has to give it up and prefer instead a competitor that is easier to test, and this means that we cannot accomplish all our methodological wishes at the same time. The Popperian cannot have his cake and eat it too.

It seems, after all, that Popperian epistemology is not exclusively fallibilist as we may have thought. As seen, it seems to rely heavily on corroboration in order to account for the progress of science. Moreover, according to Derksen, the project of achieving methodological unity by means of falsifiability collapses here. For Popper denies all inductive assumptions in corroboration, but claims that a well-corroborated theory has to be considered as a good approximation to the truth. The problem is that corroboration might not be inductive with respect to the past and the present, but it is inductive with respect to the future. Popper maintained that when a theory withstands a series of very varied and risky tests, it is 'highly improbable that this is due to an accident; highly improbable therefore that the theory is miles from the truth'.[83] And this is all that Derksen thinks he needs to support his argument, because he sees the 'whiff of induction' as the recognition that even Popperians have to accept the inductivist principle that the future will be like the past. Otherwise, the distribution of a theory's truth and falsity contents will not stay so favourable for us.

I shall deal with the criticisms of corroboration in the next chapter. Let me focus here on Derksen's claims on falsifiability. The gist of his criticism is that Popper uses the concept of falsifiability in an ambiguous way. However, Derksen does not explore any possible way of making the concept unambiguous. He is convinced that Popper needs this ambiguity plus some suppressed inductive assumptions to obtain the 'great chain' that secures the unity of his epistemology. Since Popper himself has disclosed two senses of falsifiability and has given all the alternative definitions of the term, Derksen seems to be making a fair point. However, I think his criticism is derived from some misunderstandings. To begin with, the distinction between (a) and (b) above is both erroneous and pointless. According to *Fsb*2, a theory is falsifiable if it prohibits at least one event. But test-statements that become corroborated are not (properly speaking) members of the class of potential falsifiers (PFs) of a theory. Derksen is confusing here the story of a falsifiable theory that has withstood some tests with the requirements to qualify it as falsifiable at any given moment. If T_N was falsifiable at t_0 because it prohibited event y, one cannot say that y is still in the class of potential falsifiers of T_N at time t_1, when y has been corroborated. To count T_N as falsifiable at t_1 one needs to rely on another potential falsifier. Derksen's distinction is erroneous because testability amounts to the feasibility of overthrowing a theory (if its informative content is high, then it could be easy to falsify the theory) and is never expressed in terms of the PFs that have not been tested. That the PFs have not been tested is a given (otherwise they could not be members of such a class); how easy it is to test them is another matter, but certainly not problematic in Popper's theory. Derksen's distinction is pointless, because (a) and (b), if correctly understood, are interdependent. The nature of the PFs yields both the informative content and the ease of testability of the theory. Regarding clauses (c) and (d) let me point out that they cannot be correct simultaneously. If T_N has a high degree of falsifiability, it is because its class of PFs is larger than the corresponding class of a competitor, or because it prohibits a great deal. But T_N cannot be highly falsifiable if it does not have the relevant tests left, as stated in (d). Here Derksen is confusing the two senses of falsifiability: a theory can have a high degree of falsifiability$_1$ and never become falsified, but this has been explained satisfactorily above. Finally, (e) is not very precise. Informative content and testability are ways to cash out falsifiability, but they hardly make two different concepts. Rather, they express interdependent logical requirements.

Notes

1. This point is made frequently in Popper's main works. In one of the most clear instances he writes:

   I proposed (though years elapsed before I published this proposal) that the *refutability or falsifiability* of a theoretical system should be taken as the criterion of its demarcation. According to this view, which I still uphold, a system is to be considered as scientific only if it makes assertions which may clash with observations; and a system is, in fact, tested by attempts to produce such clashes, that is to say by attempts to refute it. Thus, testability is the same as refutability, and can therefore likewise be taken as a criterion of demarcation. (*CR*, p. 256.)

2. This is the crudest formulation of falsifiability and, if taken literally, can lead to misunderstandings. The reader should be patient until the notion is fully fleshed

Falsifiability
75

out. On the other hand, although theories (set of statements) and states of affairs (experience) are not the sort of entities that can 'clash', this way of talking is common parlance and taken as a shorthand for 'theories clash with the statements that describe states of affairs'. I see no special reason to deviate from this usage and therefore will stick to the expression.

3. Popper 1968, p. 139.
4. *LScD*, p. 33.
5. Not all cases of theory replacement involve falsification. Within the premises of falsificationism, one can reject a non-falsified theory in the light of a better-fitted competitor.
6. Strictly speaking, there is also the possibility that a theory may subsequently fail the *same* test that it has passed before. This, however, is usually attributed to defects in the testing process (e.g. carelessness of the experimenter, inaccuracies in the measures, corrupted data) rather than to arbitrary changes in the state of affairs. On the other hand, one could say also that a negative decision is not definitive either, since no falsification is conclusive (cf. *RAS*, p. xxi). I shall give more details about the temporary character of corroboration in Chapter 4.
7. *LScD*, p. 34. Cf. also p. 30 (where Popper draws a contrast between inductivism and deductivism), section 3, pp. 32–4, section 5, p. 39, section 85, pp. 276–8, and the appendixes in pp. 315–16, 412–13. Popper does not limit himself to refuting the naïve version of verification. As I will show in Chapter 4, he has well-developed arguments against inductive probability.
8. Cf. ibid., p. 47.
9. Ibid., p. 41 (parentheses suppressed).
10. See Hempel 1992, pp. 71–84. Incidentally, Hempel's version of falsifiability is both inaccurate and misleading. It is inaccurate because it conflates Ayer's second definition of 'verifiability' with a simplistic version of falsifiability; it is misleading because it makes the falsificationist look like someone who is driven by the search for a criterion of meaning, a motivation which at least in Popper's case is completely misconceived. Joseph Agassi makes some interesting comments on the positivist reading of the theory of falsifiability (see Agassi 1993, pp. 98ff.).
11. The broad line of demarcation between empirical science on the one hand, and pseudo-science or metaphysics or logic or pure mathematics on the other, has to be drawn right through the very heart of the region of sense – with meaningful theories on both sides of the dividing line – rather than between the regions of sense and of nonsense. I reject, more especially, the dogma that metaphysics must be meaningless. For as we have seen, some theories, such as atomism, were for a long time non-testable and irrefutable (and, incidentally, non-verifiable also) and thus far 'metaphysical'. But later they became part of physical science. (*RAS*, pp. 175–6)

    See also *LScD*, pp. 37ff. and *RAS*, p. xix. For a detailed discussion of Popper's views on the relationship between metaphysics and science see Agassi 1964.
12. 'Purely existential statements are not falsifiable – as in Rudolf Carnap's famous example: "There is a color ('Trumpet-red') which incites terror in those who look

at it"' (*RAS*, p. xx). In another example Popper writes: 'There exists (somewhere in the universe, either in the past, present or future) a raven which is not black.' This is non-testable, since it cannot be falsified by any amount of observation reports (*RAS*, p. 178). The common reader would surely consider the thesis of the assimilation of existential statements with metaphysical statements as odd, for it is difficult to see how anyone could treat as metaphysical a statement that describes directly verifiable facts, and concerning which it is possible to decide without major difficulties. Popper's view becomes meaningful when we remind ourselves that many existential statements are excessively restrictive and, in general, only provide us with trivial information.

13. This is so because the main purpose of the demarcation criterion is the elimination of non-falsifiable statements (metaphysical elements) from theoretical systems (to increase their testability) instead of the elimination of meaningless statements.
14. *LScD*, p. 69.
15. *RAS*, pp. 178–9; see also *LScD*, p. 70 and Schilpp (ed.) 1974, pp. 202ff.
16. Thus one can certainly say that falsifiability and verifiability are 'symmetrical' in the sense that the negation of a falsifiable statement must be verifiable, and *vice versa*. But this fact, which I repeatedly stressed in my *LScD* . . . and which is all that my critic establishes in his premises, is no argument against the fundamental asymmetry whose existence I have pointed out. (*RAS*, pp. 183–4)
17. Ibid., p. 183. Again, the asymmetry between verification and falsification springs from the fact that a singular statement or 'a finite set of basic statements, *if true*, may falsify a universal law; whereas, *under no condition* could it verify a universal law: there exists a condition wherein it could falsify a general law, but there exists no condition wherein it could verify a general law' (p. 185). For a clear explanation of this relation see Magee 1973, p. 15.
18. *LScD*, p. 86. Cf. also section 6, pp. 40–43 (see Chapter 4).
19. Cf. ibid., p. 84. It seems that Popper considers the class of factual basic statements as a big subclass of all possible statements that would include also those statements that are completely unrelated to empirical science, namely strict metaphysic statements and many more.
20. The way in which such increase can be obtained is itself problematic (compare increase as a result of tautological operations – where the newly deducted statements have less empirical content that their premises – with increase through non-apodictic means). I will, however, set this point aside temporarily.
21. According to *Fsb*1 only one prohibited statement is required. This has misled many authors into thinking that in Popper's philosophy a single existential statement (and even sometimes a single observation) suffices to falsify a theory. (See, for example, Johansson 1975; Narayan 1990.) That this is a mistake can be easily shown in two ways: (1) the requirement of intersubjectivity of falsification precludes the possibility that a single observation might be sufficient; (2) Popper's treatment of isolated existential statements and his contention that falsifying statements should be, at their turn, testable, rule out the vast majority of existential statements. Thus, the availability of an existential statement is a necessary

Falsifiability 77

but not a sufficient condition for falsifiability. As I shall show in my comments to *Fsb*2, some further qualifications are needed.

22. Popper distinguishes two senses of the term 'falsifiable'. Let me use subscripts to refer to each of them. 'Falsifiable$_1$', is a technical logical term (sometimes expressed by the locution 'falsifiable in principle') that 'rests on a logical relation between the theory in question and the class of basic statements (or the potential falsifiers described by them)'. 'Falsifiable$_2$' refers to the possibility that a theory is actually refuted by experience. (See *RAS*, p. xxii; *LScD*, sections 4 and 22.) As is obvious, the second sense is stronger than the first, for any theory that is falsifiable$_2$, is falsifiable$_1$, though the converse does not hold. Since the second sense of falsifiability can be perfectly conveyed by the word 'falsification', is not Popper's distinction innocuous? According to him, one should aim to develop theories so risky that they cannot escape falsifiability$_2$, for this promotes the growth of science through the rejection of unfit alternatives and the subsequent formulation of better (or purportedly better) competitors. But the weaker sense of falsifiability proves to be enough to keep Popper's method running. Just falsifiabilty$_1$ needs to be used to filter theories, characterize them as empirical and choose the better ones. However, many criticisms of Popper's theory of science fail to distinguish between falsifiability$_1$, the mere compliance of formal conditions imposed on the statements of a scientific theory and falsifiability$_2$, the eventual realization of the clash. As to the latter, which is obviously more troublesome, many have argued that the provisional character of every falsification proves Popper wrong and discloses the redundancy of the second sense. I think this is a misunderstanding. Strictly speaking, one could mention varied instances of 'definite' falsification that block such objections. Further, many criticisms are supposed to target the weaker sense of the notion but focus all their arguments against the stronger version. So, though whoever makes a sharp distinction between *falsifiability* and *falsification* can do perfectly well without Popper's two senses, I think it is worth keeping the distinction in mind in order to deal with many texts that address falsifiability.
23. Cf. *LScD*, p. 86.
24. This, however, does not preclude us from finding actual instances of falsification that can be agreed upon without jeopardizing the use of the notion: 'there are a number of important falsifications which are as "definitive" as general human fallibility permits. Moreover, every falsification may, in its turn, be tested again' (*RAS*, p. xxiii). On the other hand, as Ackermann rightly notes, actual falsification need not lead to rejection especially when the falsified theory 'is still in close enough approximation to the data to be useful in solving problems and no alternative nonfalsified theory has as yet been constructed to replace the falsified theory' (Ackermann 1976, p. 18).
25. All that matters in *Fsb*1 is the relationship between the system of statements (the theory) and the class of its potential falsifiers. The latter is supposed to be comprised of basic statements, and in view of complying with the restriction of not being an empty class, it is enough if there is at least one basic statement in it, no matter the characteristics of this statement, nor its particular or general character, nor the mechanism adopted to obtain it.

26. See *LScD*, section 23. However, shortly after this amendment, Popper introduces one more change and goes from 'occurrences' to 'events' in the full-blown definition of $Fsb2$.
27. Ibid., p. 89.
28. Speaking of the singular statement p_k, which represents one occurrence P_k, one may say, in the realistic mode of speech, that this statement asserts the occurrence of the event (P) at the spatiotemporal position k. And we take this to mean the same as 'the class P_k, of the singular statements equivalent to p_k, is an element of the event (P)'. (Ibid.)
29. $Fsb1$ was also insufficient on another respect. The criterion is too loose if one does not restrict: (a) any possibility of modifying at will the number of members in the class of potential falsifiers by performing apodictic operations and (b) the characterization of a non-empty class of potential falsifiers by the fact that it contains a basic statement that describes an existential fact. The latter restriction is particularly critical since a system of statements that prohibits a purely existential fact in some sense prohibits very little. Besides, if we take into account that by definition, most empirical existential statements are not falsifiable, failure to restrict (b) would be tantamount to characterizing a class of potential falsifiers appealing to statements that are not falsifiable themselves, which goes against a previously introduced caveat.
30. Ibid., p. 90.
31. Ibid., p. 113; see also *CR*, p. 385; *RAS*, pp. 239, 245–6, 249.
32. It is worth mentioning that Popper himself never uses the term 'falsificationism' (see *RAS*, p. xxxi) among other reasons because he has refused to turn 'falsifiability' into an 'ism' and has resisted making his proposal into a doctrine. Instead, he uses 'falsification', 'falsifiable' and all its variants.
33. Popper started using the locution 'test-statement' in his 'Replies to my critics' (Popper 1974). This locution refers to what was called 'basic statements' in *LScD*.
34. In what follows I shall presume that the reader has a minimal familiarity with Kuhn's image of science, hence I will not explain his views on this matter.
35. Kuhn, 1962, p. 77.
36. For a time Popper shared Kuhn's enthusiasm about their coincidences (see his contribution to *Criticism and the Growth of Knowledge*). Later on, Popper considered it necessary to point out the differences with respect to Kuhn's views of science. On the other hand, Kuhn mentions ten points of agreement and concludes that 'though [that list] by no means exhausts the issues about which Sir Karl and I agree, is already extensive enough to place us in the same minority among contemporary philosophers of science'. See Kuhn 1974, p. 2 (footnote suppressed).
37. There is a sense in which Kuhn supports scientific realism, though it should be clear that it is not the kind of scientific realism that Popper endorses. The following passage is one of the few places in which Kuhn clearly gestures towards realism: 'we both [Popper and Kuhn] insist that scientists may properly aim to invent theories that *explain* observed phenomena and that do so in terms of *real* objects, whatever the latter phrase may mean' (ibid.).

38. Ibid.
39. Briefly stated: 'though tests occur frequently in normal science it is the individual scientist rather than the current theory which is tested' (ibid., p. 5).
40. Cf. ibid., p. 14. That would keep Popper committed to falsification as conclusive disproof.
41. Ibid., p. 15.
42. Ibid., p. 22. The Popper–Kuhn debate can be cast along the two dimensions announced in the title of Kuhn's paper: logic vs psychology, science vs scientists, principles vs institutions and theory vs reality. However, as suggested above, those dimensions need not be in conflict but might be complementary. We can see that this is a feasible project taking as an example the way formal theory and pragmatics complement each other in philosophy of language. *Mutatis mutandis*, in scientific research we can strive for something alike and retain falsifiability (a logical principle) as the core of the testing process while construing the pragmatics according to what Kuhn's theory has elucidated. Global changes may depend more on the pragmatics, and yet the principle of each local decision as to whether a hypothesis is true or not is still determined by the logical processes of falsifiability (unless corruption among the participants is present).
43. The term 'puzzle' is meant to stress that 'the difficulties which *ordinarily* confront even the very best scientists are, like crossword puzzles or chess puzzles, challenges only to his ingenuity. *He* is in difficulty, not current theory' (ibid., p. 5, n. 1).
44. Popper, 'Normal science and its dangers', in: Lakatos and Musgrave (eds) 1970, p. 52.
45. Ibid., p. 57. For a more articulated discussion of the myth of the framework see *MF*, pp. 33–64
46. Ibid., p. xxxv. Only naïve falsificationism requires infallible test-statements. By contrast, the fallible character of test-statements (which – in Popper's theory – is a consequence of the fallibility that may affect any step of scientific activity) seems to be needed under any minimally sophisticated interpretation of falsifiability.
47. Gupta thinks that Putnam's 'internal realism' settles the Popper–Kuhn debate. See Gupta 1993.
48. Kuhn seems to have trouble understanding that fallibilism (about knowledge and test-statements) is perfectly consistent both with the definitive standards of a stringent criterion of demarcation and with the rationality of decisions about testing. As D'Amico nicely puts it: '[f]allibilism shows that dogmatism or certainty about knowledge is irrational, but does not lead to total or self-defeating scepticism because of the ability to disprove theories by logical refutation' (D'Amico 1989, p. 25).
49. '"What kind of observation would refute to the satisfaction of the Newtonian not merely a particular Newtonian explanation but Newtonian dynamics and gravitational theory itself?" The Newtonian will, alas, scarcely be able to give a positive answer' (Lakatos 1974, p. 247). I am afraid that Lakatos's implicit conclusion overstates the normative component in Popper's theory of science.

80   *Popper's Theory of Science*

Mutatis mutandis, a similar question can be posed to any philosophy of science obtaining comparable results, but difficulties to fit exactly a particular methodological rule hardly prove its inadequacy.

50. Lakatos offers what he takes is a more suitable alternative: 'The best opening gambit is not a falsifiable (and therefore consistent) hypothesis, but a research programme. Mere "falsification" (in Popper's sense) must not imply rejection. Mere "falsifications" (that is, anomalies) are to be recorded but need not be acted upon' (Lakatos 1981, p. 116, references suppressed).
51. Lakatos 1970, p. 99.
52. Cf. ibid., p. 102. In the absence of conclusive disproof and without a clear-cut distinction between theory and empirical basis, the dogmatic-falsificationist criterion of demarcation breaks down. Now, if a refuting instance is interpreted as indicating that there may be another cause operating simultaneously, tenacity of a theory against empirical evidence would become a plus, for in Lakatos's view, it could foster the development of the theory in new fields. This conclusion, however, appears to be mistaken. If the aim of science is the search of truth – as Popper maintains – it is hard to see how refusing to accept refutations by way of modifying the *ceteris paribus* clause might give us an appropriate *succedaneum* for this aim. In other words, one may suspect that a theory that faces non-problematic counter-instances is false, hence recommending tenacity against empirical evidence (by calling it a desirable feature of a theory) does not get us any closer to the truth.
53. Ibid., p. 103.
54. Cf. ibid., pp. 106–7.
55. If the conventionalist decisions of the methodological falsificationist are made as Lakatos represents them, they are in fact too arbitrary. But Lakatos has a recipe to avoid this flaw and overcome the shortsighted idea that a test is always a confrontation between theory and experiment. Since a competing theory must enter the testing scenario, basic statements should be accepted in accordance with a 'research programme', that is, in accordance with certain prearranged theoretical considerations. This takes us to 'sophisticated falsificationism', which differs from naïve dogmatic falsificationism: 'both in its rules of acceptance (or 'demarcation criterion') and in its rules of falsification and elimination. For the naïve falsificationist, any theory which can be interpreted as experimentally falsifiable is 'acceptable' or 'scientific'. For the sophisticated falsificationist, a theory is 'acceptable' or 'scientific' only if it has corroborated excess empirical content over its predecessor (or rival): that is, only if it leads to the discovery of novel facts. See ibid., p. 116 (footnote suppressed). Cf. also 'Lectures on scientific method', in Motterlini 1999, p. 104.
56. Cf. Duhem, 1954, Chapter 6, sections 1 and 10.
57. Lakatos formulates his response to the question, How does science progress?, in terms of the notion of research programmes in which an amended criterion of demarcation may solve the problems of unfit varieties of falsificationism. In his opinion, this formulation represents a shift from the aim of naïve forms of falsificationism (the appraisal of a theory) to the problem of how to appraise *series* of

theories. The acceptance of basic statements as conventions may be extended to universal statements to foster scientific growth. See Lakatos 1974, p. 248.

58. Cf. Galileo, 'Letters on sunspots' in Drake (ed.), 1957, pp. 59–144. Incidentally, the Italian scientist does not make a big deal of the refutation of the Aristotelian doctrine of the immutability of the heavens. Galileo even speculates that had Aristotle been in the possession of some sensory evidence on phenomena like the sunspots, he would have formed a different opinion on the nature of the heavens, closer to Galileo's own views.

59. Galileo 1961, p. 11. The discussion on the reliability of instrument-aided observation as compared to natural vision has a long history. For an illuminating treatment see Steneck and Lindberg 1983.

60. The reader must be aware that Popper endorses a weaker claim about the theoretical/observational distinction, according to which observation is theory-laden. This, however, should not be confused with my criticism of Lakatos's treatment of such distinction. My point (that Galileo's case does not support Lakatos's argument) still holds.

61. Cf. ibid., p. 131. Interestingly enough, four pages earlier he wrote:

> [w]e cannot avoid the decision which sort of propositions should be the 'observational' ones and which the 'theoretical' ones. We cannot avoid either the decision about the truth-value of some 'observational propositions'. These decisions are vital for the decision whether a problem shift is empirically progressive or degenerating.

62. Cf. ibid., p. 116. The fitness of the competitor, say t_2, is determined by the satisfaction of three requirements: (1) t_2 has more empirical content than t_1; (2) the empirical (unrefuted) content of t_1 is a proper subclass of the empirical content of t_2; (3) some of the excess content of t_2 is corroborated. None of these requirements is an original contribution of Lakatos's, they all appear in Popper's later explanation of the process of theory selection and replacement.

63. This accusation is a consequence of the more general charge that Popper conflates refutation and rejection. In his 'Replies to my critics' (Popper 1974), Popper admits that he has talked about the latter when discussing the former due to the needs of argumentation, but he explains carefully that falsification is a logical matter, whereas the rejection of a theory and the necessity to abandon it is a question of methodology. What Popper really means is that a refuted theory has to be rejected as a contender for the truth, not that it has to be abandoned as soon as is refuted.

64. For the close affinities between Feyerabend and Lakatos see their correspondence in Motterlini 1999.

65. This shortcoming affects falsification. In particular, 'a strict principle of falsification or "naïve falsificationism", as Lakatos calls it, would wipe out science as we know it and would never have permitted it to start' (Feyerabend 1975, p. 176, footnote suppressed).

66. Ibid., p. 171.

67. Ibid., p. 66.

68. It is assumed that test-statements involved in 'arguments from observation' are firmly connected with appearances, but this is not the case: '*theories* ... which are not formulated explicitly enter the debate in the guise of observational terms' (ibid., p. 75; parentheses suppressed).
69. Ibid., p. 76. Cf. Popper's frequent pronouncement about the impossibility of making observations without a proper conceptual framework.
70. Ibid., p. 179.
71. For a thorough discussion of this see D'Amico 1989, pp. 41ff.
72. Cf. *RAS*, p. xxvi.
73. Feyerabend's interpretation of many episodes of the history of science has been disputed on textual evidence. Noticeable is his debate concerning the meaning of the so-called tower experiment (Galileo's thought experiments on falling bodies) with Machamer. I deal with Galileo's work at some length in my *Historical Evolution of Scientific Thought* (García 1998).
74. The book was published on 12 March 1610 (a few days after Galileo's last reported observation, so eager he was to communicate his discoveries!). See Galileo 1961.
75. By contrast, there were many common-sense observations that supported the Ptolemaic system.
76. There were also amateurs who not only made numerous astronomical observations consistent with the ones reported by Galileo, but were able to use them for practical purposes, such as solving the problem of how to make an accurate estimate of longitude. The French courtesan de Peiresc is one such case. He used his observations of Jupiter's satellites to prepare tables which would permit observers stationed in widely separated areas to compare their configurations of Jupiter's moons and deduce from them the difference of time and hence longitude. The fact that this project started as early as November 1610 supports the contention that Galileo's observations were quickly validated. For the whole story see Chapin 1957. I am indebted to Robert Hatch for having called this rare reference to my attention.
77. Derksen 1985, p. 313.
78. Ibid., pp. 314, 315, 321. Derksen motivates this claim as an important component of falsifiability invoking the relationship between testability and corroboration. He points out that since the best testable theory offers the best chance to learn something, the best-corroborated theory (where falsification has not been the outcome of testing) provides some reasons to believe that we have come closer to the truth. Of course, Popper would replace the word 'believe' in Derksen's Claim III by the word 'guess' or 'conjecture'.
79. Ibid., p. 318. Popper's quotation comes from *CR*, p. 263. I shall deal with Derksen's qualms about corroboration in the next chapter. For the time being, let me point out that Derksen's remark about a series of refuted theories of increasing falsifiability is a bit confusing. The elements of a sequence of refuted theories (once refuted) have the same degree of testability. What one may argue is that they are formulated in such a way that each subsequent theory is easier to falsify than its predecessor, so that, *before* falsification, each member of the series has an

Falsifiability

increasing degree of falsifiability. However, there are hardly examples of such a series – assuming Derksen has in mind a series of theories where each member is a modification of a previous element of the set, instead of just a bunch of false theories put together in a series.

80. Ibid., p. 320.
81. Ibid., p. 323 (brackets added).
82. Ibid., p. 324.
83. Popper 1974, p. 1041.

4
Corroboration

'Theories are not verifiable but they can be "corroborated".' This is the opening sentence of Chapter 10 of *LScD* and expresses, in a very condensed way, the sharp contrast that Popper wants to draw between his approach to science and any verificationist approach. As discussed in the previous chapter, falsifiable theories make negative predictions of particular states of affairs. In this sense, if such predictions fail to be realized, then the theories become confirmed, well-tested or corroborated.[1] Corroboration only shows how well a theory has stood up to the tests up to a certain moment of time; it does not indicate how well the theory will fare in the future, since it might fail the very next test. We tend to favour those theories that are well-corroborated since they have proved capable of giving a good explanation of a certain sector of reality and they have shown that they do not conflict with experience in those areas in which the prohibited fact did not take place. Besides, when we have to choose between competing theories, we prefer those that are better corroborated, provided they are also testable.[2] In this chapter I explain Popper's notion of corroboration, examine the most important objections formulated against this notion and dispense with the charge that corroboration is bound to presuppose induction by showing that this is a misconstrual of Popper's proposal.

Popper's initial characterization of this notion in *LScD* had a twofold motivation. Firstly, Popper wanted to avoid, as much as possible, any discussion of theories or hypotheses in terms of truth and falsity because, by this time, he regarded the notion of truth as problematic and considered that he could formulate his account of science without appealing to it.[3] He was looking for a different notion to describe the degree to which a theory or a hypothesis has survived a test. Furthermore, he wanted a notion that could afford him his non-justificationist stance, a notion that was not 'ultimate'. Secondly, as a result of his rejection of induction, he wanted to discredit the popular idea that theories (or hypotheses), though they could not be proved true, could nevertheless be shown to be more or less probable. He was against 'probability logic' or the ascription of degrees of justification to scientific statements. In particular, he thought that instead of discussing the probability of a hypothesis we should be concerned with assessing its performance in the testing situation. Let us recast Popper's arguments against the attempt to capture the notion of corroboration with the assistance of the theory of probability.[4]

Corroboration

4.1 Corroboration and probability of events

Some supporters of induction have turned their attention to the frequency theory of the probability of events as a means of avoiding the difficulties of verification.[5] To put it briefly, they claimed that though it is not possible to establish any (universal) statement with certainty, it is possible to determine its degree of probability within tolerable margins of error, which achieves a twofold objective: (1) enabling one to choose from alternatives of different degrees of certainty; and (2) making sense of a quantitative approach to certainty (and perhaps finding a level of probability that is the appropriate surrogate for certainty). Moreover, they think that the probability of hypotheses can be reduced to the probability of events, which can be expressed in its turn in terms of the probability of statements.[6] On this line of thinking, they regard the probability of a hypothesis as a special case of the probability of a statement. According to the frequentist interpretation of probability, $p(x) = \#OFO/\#OO$ (where 'x' is an event, 'OFO' means observed favourable outcomes, and 'OO' means observed outcomes). The verificationist extends this treatment to the probability of hypotheses, and holds that in a finite sequence $p(h) = \#C/\#W$ (where 'h' is a given hypothesis, 'C' stands for confirmed instances and 'W' stands for the total number of instances).

Popper has two worries concerning this view. The first is that the theory of the probability of hypotheses seems to be the result of the confusion of psychological and logical questions. Popper does not wish to deny that one may have subjective feelings of different intensities with regard to the degree of confidence with which one expects the fulfilment of a prediction (and the consequent corroboration of a hypothesis) in the light of the past performance of a theory. But as discussed in Chapter 2, these feelings belong to the psychology of research and are foreign to epistemology. The second worry is whether the identification of the probability of hypotheses with the probability of statements – and indirectly with the probability of events – is a correct move.[7] Here Popper sees another confusion: that assimilation is unwarranted and one should never use the expression 'probability of statements' when discussing the probability of events. Besides, all attempts to carry out that identification are doomed to fail. The core of his argument is that under no circumstances can any statement attributing probability to a hypothesis be translated into a statement about the probability of events. Three reasons support this contention.

1. Attempts to reduce the idea of a probability of hypotheses to that of a truth-frequency within a sequence of statements (in accordance to Reichenbach's suggestion) lead to undesirable consequences. One way to interpret Reichenbach's view involves taking the various singular statements that can

agree with or contradict a hypothesis as the elements of its sequence. Let h be the hypothesis 'all ravens are black' to which we are attributing probability and $k_1 \ldots k_n$ the sequence of statements that we use to determine the truth frequency. Assume that h is refuted, on the average, by every second single singular statement of the sequence (so that $k_1 = 1, k_2 = 0, k_3 = 1, k_4 = 0 \ldots$). Then this would force us to attribute a probability of $\frac{1}{2}$ to h. On the other hand, suppose that we appraise the probability of h by using the ratio of the passed tests to the tests that have not been attempted yet. If we treat the second class as an infinite class, we would have to attribute a probability of zero to h, even if it has passed all of the tests attempted so far. Consider, for a change, that the results of the tests to which h has been subjected include confirming (ct) as well as indifferent (dt) instances. How should we determine the probability of h? If we use the ratio ct to dt, and treat dt as false, we would be doing an unwarranted assimilation. Moreover, if by an indifferent result we understand one which did not produce a clear decision, we face the risk of letting subjective considerations about the skill of the experimenter leap into the appraisal of h.

2. The type of hypothesis in which we are interested in science cannot be a sequence of statements, for a strictly universal statement is not equivalent to a long conjunction of singular statements. Hence, the suggestion that a hypothesis can be treated as a sequence of statements (for the purpose of attributing probability to it) is misconceived. In addition, basic statements are not derivable from universal statements alone, whereas its negations are. Assume we take the sequence of these negations and make it equivalent to a given hypothesis. In this case, the probability of every self-consistent hypothesis will be 1. For we need to use the ratio of the nonfalsified (derivable) negated basic statements (which together with the class of the derivable statements are infinite classes) to the falsified ones (the accepted falsifying basic statements), which at most constitute a finite number.

3. Just as we might consider a singular occurrence 'probable' if it is part of a sequence of occurrences with a certain probability, we might call a hypothesis 'probable' if it is an element of a sequence of hypotheses with a definite truth-frequency. But this does not work, for attribution of probability presupposes that we ignore whether a given hypothesis is true (otherwise we would not need the concept of probability). Accordingly, we would not be able to say which hypothesis is true and which is false in a given series in order to establish the truth-frequency required for the estimative of probability. If we try to overcome this result by turning to the falsity-frequency within a series as our starting-point, then using the ratio of falsified to non-falsified hypotheses, and taking the reference sequence as infinite will make the probability of every hypothesis in such a sequence equal to 1, as pointed out in (2). If, on the other hand, we treat the sequence as finite and ascribe to each of its elements a

Corroboration 87

probability between 0 and 1, then every falsified hypothesis (fh) (in so far as it belongs to the sequence) should be accorded a positive degree of probability (that would decrease, with every newly fh, according to the ratio fh/n, where n is the total number of hypotheses in the reference sequence). In sum, attributing a positive degree of probability to a false hypothesis seems to contradict the attempt to express in terms of 'probability' the degree of reliability that we want to ascribe to a hypothesis in view of supporting or undermining evidence.[8]

On Popper's view, the project of constructing a concept of the probability of a hypothesis that intends to capture its 'degree of certainty' in analogy with the concepts 'true' and 'false' is hopeless. However, the supporters of inductive logic rejoin that one can attribute a definite sense of probability (with specification of a numerical degree or not) to a scientific hypothesis by means of a perfectly understandable statement. This statement (e.g. 'the law of universal gravitation is highly probable') that asserts the appraisal of such hypothesis, also tells us up to what degree it is adequate. But analysis of the logical status of this kind of statement brings in again the problem of induction. The statement is obviously synthetic, but it is not verifiable since the probability of a hypothesis cannot be conclusively deduced from any finite set of basic statements. Furthermore, it can be neither justified nor tested. And when we turn to the appraisal itself and try to determine whether it is true or probable, we quickly get in trouble. We can consider it true only if we take it as an *a priori* true synthetic statement (since it cannot be empirically verified). If we consider it probable, then we would run into an infinite regress and would need to move from the appraisal of the appraisal to an appraisal of higher level.

Since the corroboration of a theory can only be expressed as an appraisal (exactly like the notion of probability just examined), it seems that Popper's objections to probability logic, and especially the charge that it brings back the pitfalls of induction, can be turned against his own views. But this is erroneous on two counts. First, the appraisal of hypotheses that describes them as 'provisional conjectures' and denies that they can be asserted as 'true' has the status of a tautology, because it simply paraphrases the claim that strictly universal statements cannot be derived from singular statements. This means that the second objection to probability logic cannot be raised, and the first objection would be irrelevant because statements asserting corroboration appraisals are not hypotheses, though they can be derived from the theory plus the accepted basic statements. The appraisal made by means of corroboration statements simply states that the accepted basic statements 'do not contradict the theory, and it does this with due regard to the degree of testability of the theory, and to the severity of the tests to which the theory has been subjected, up to a stated period of time'.[9]

4.2 The notion of corroboration

As has been adumbrated in the foregoing, we call a theory 'corroborated' if it has stood up to some empirical tests. Now we need to spell out this notion. To begin with, we should refrain from treating corroboration as a purely logical affair.[10] Although the corroborating appraisal establishes relations of compatibility and incompatibility between the theory and a set of test-statements, we cannot attribute a positive degree of corroboration to a non-falsified theory only because it is compatible with a given system of known basic statements. Here actual incompatibility is sufficient to declare falsification, but mere compatibility is not sufficient to accord a positive degree of corroboration to a theory, since even metaphysical systems can be formulated in such a way that they become compatible with a system of accepted basic statements. Secondly, we need to cash out corroboration as a comparative notion that enables us to rank theories according to how well they pass the tests to which we submit them. We are interested in the severity of the tests and the ability of a theory to pass them.[11] Accordingly, we qualify theories as better or less well corroborated than their competitors. Thus, we check that a theory has been tested and that it turns out compatible with a non-empty subclass of accepted basic statements. It may be tempting to ascribe a significant degree of corroboration to a theory that is compatible with many test-statements. However, this is misleading in two ways. (1) The degree of corroboration cannot be determined by merely counting the number of compatible test-statements, for the *nature* of the tests has to receive the main weight in the appraisal (tests can also be ranked according to their degree of *severity*). (2) For a pair of non-competing hypotheses α and β we may want to accord a higher degree of corroboration to β, even if it has passed fewer tests than α. For suppose α is the hypothesis 'all ravens are black', while β is Einstein's hypothesis about the proportionality of inert and (passively) heavy mass. Although α might have more corroborating basic statements, we certainly consider Einstein's hypothesis to be best corroborated. This caveat about the role of the number of corroborating instances illustrates that what really matters for determining the degree of corroboration is

> the *severity of the various tests* to which the hypothesis in question can be, and has been, subjected. But the severity of the tests, in its turn, depends upon the *degree of testability*, and thus upon the simplicity of the hypothesis: the hypothesis which is falsifiable in a higher degree, or the simpler hypothesis, is also the one which is corroborable in a higher degree.[12]

I want to call the reader's attention to the last sentence in the previous quotation. Popper is not establishing a parallel between falsifiability and corroboration, but between falsifiability and *corroborability*. And the case considered represents just an ideal situation. When a test is performed and the attempted falsification does not obtain, the results may go in ways that preclude comparison and attribution of degrees (to both falsifiability and corroboration) in the simple direct way that the quotation suggests. One and the same hypothesis may have a high degree of falsifiability and be corroborated only slightly; or it may become falsified; or even it may turn out to be very well corroborated but lose the features that determined its high degree of falsifiability. As in the characterization of falsifiability, it is easier to define a numerically calculable degree of corroboration for instances that are in (or close to) the extremes. Popper formulates some rules to help make attributions of positive or negative degrees of corroboration. For example, we should not continue to attribute a positive degree of corroboration to a theory that has been falsified by an intersubjectively testable experiment.[13] Furthermore, the addition of new statements to the set of accepted basic statements and the performance of new tests might change a corroborative appraisal, replacing a positive degree of corroboration by a negative one (but not the reverse, unless there has been a demonstrable mistake in the first attribution).[14]

4.3   Corroboration and logical probability

Popper stipulated values for the limiting cases, assigning 0 to tautologies and 1 to self-contradictory statements. This assignment, which is done only by virtue of the logical form of the statement, expresses the connection between degrees of falsifiability and his concept of 'logical probability'.[15] Popper equates 'logical probability' to 'inversed testability' and stresses that the former is inversely proportional to empirical content. In other words, falsifiability and logical probability are inversely related in such a way that to the maximum degree of falsifiability corresponds the minimum degree of logical probability and vice versa. Thus, the best testable statement is the one with the lowest logical probability. Since we have already established a relation between testability and corroborability, it turns out that the best-testable statement will also be the better corroborable and the most logically improbable (or the less logically probable). This result shows also why we cannot treat corroboration with the machinery that is used to treat the probability of events and that calculates this measure in a direct proportion to their logical probability.

Another reason that precludes the identification of corroboration with logical probability can be stated in the following way. To make appraisals of corroboration for a given theory we take into account the previously assigned degree of testability. For this purpose we assign better marks to theories that are logically improbable and survive the tests. On the other hand, the grading can be proportional to the number of corroborating instances, as long as it is always respectful of the following caveat: high degree of corroboration will go hand-in-hand with a positive outcome *only* for the first corroborating instances, since 'once a theory is well corroborated, further instances raise its degree of corroboration only very little'.[16] Nevertheless, if subsequent positive outcomes corroborate the theory in a *new* field of application, then we do not have to abide by this rule. In such a case, we can increase considerably the degree of corroboration accorded to the theory giving due credit to the new instances. According to Popper, the degree of corroboration assigned to a theory which has a higher degree of universality usually surpasses the degree of corroboration assigned to a more restricted theory.

Note that corroboration is not a truth-value; hence, one cannot use the notion of corroboration in the same unrelativized way one uses the notion of truth, let alone define 'true' as 'corroborated', in the way pragmatists suggest. The source of this mistake seems to be the following characteristic of the logical property expressed by corroborative appraisals: an appraisal of corroboration simply asserts the logical relationship between a theoretical system and a system of accepted basic statements. As a natural consequence, it is a timeless notion – just as the concepts 'true' and 'false' (e.g. when we say that statement p is true we mean that it is true once and for all). But corroboration is not a truth-value. The main difference is that one cannot say that a statement is corroborated without mentioning the specific system of basic statements that supports the given appraisal and which is accepted at a particular time. One needs to index every corroborative appraisal to the particular system of basic statements or perhaps to the date of its acceptance, for 'the corroboration which a theory has received up to yesterday' is *logically not identical* with 'the corroboration which a theory has received up to today'.[17] This relativization to time (and to the accepted system of basic statements) makes possible that one and the same theory might receive any number of different corroboration values, according to its performance in tests attempted at various times.

Due to the complications introduced by the particularities of different test-situations (that differ regarding the accepted system of basic statements), the characteristic just sketched might hinder the use of the attributed degree of corroboration to select between theories that are well corroborated, for the epistemologist may become entangled in the multiple appraisals of corroboration with no way to make up his mind (he might not know, for example, whether he should confer more importance on one appraisal or other). For the case of competing theories, however, Popper gives us a recipe: 'a theory which has been well corroborated can only be superseded by one of a higher level of universality; that is, by a theory which is better testable and which, in addition, *contains* the old, well corroborated theory – or at least a good approximation to it'.[18]

This brings falsifiability back into the picture. It also introduces the requirement of an increase in universality, hence greater explanatory power and empirical content as part of the conditions that need to be satisfied for the replacement of theories. By demanding that the superseding theory 'contain' the older (superseded) theory, Popper secures deductive relationships between the latter and the former while describing the advancement of science as a progressive movement from theories of lesser degree of universality to theories of higher degree of universality. On the other hand, he reminds us that all scientific statements remain tentative conjectures, and this applies no less to corroborating appraisals than to the members of the system of accepted basic statements or to scientific hypotheses. Although corroboration is the outcome of failed attempts to falsify, and gives indication that our theories are on the right track, the progress of science in the long run is the exclusive result of submitting our conjectures to the most sincere criticism and the most severe tests that we can possibly formulate. Admittedly, the requirement of sincerity cannot be formalized, but subsequent criticism might correct test-situations in which it has not been exercised properly, by running the tests again and showing that they could have been made more severe.[19]

In the new appendixes, added to the English translation of *Logik der Forschung*, Popper recasts the essentials of his views on corroboration. He emphasizes again that science aims not for high probability but for high informative content backed up by experience. By contrast the most (logically) probable[20] statements are, in general, very uninformative. They tell us nothing (like tautologies) or very little, like the predictions of the palmists or soothsayers. So, the epistemologist who prefers highly probable theories faces the risk of taking as 'good' theories of very low informative content. With this in mind, suppose that we represent the measure of the severity of the tests to which a theory has been subjected with the formula $C(h, e)$ – the degree of corroboration of the hypothesis h, in the light of test e. This measure has a bearing on the question of whether we should accept (or choose) h (though tentatively). It seems, *prima facie*, that we can read $C(h, e)$ like $P(x, y)$, which asserts the relative probability of x given evidence y. But the first formula is adequate only if e expresses the report of the most severe tests we have been able to design and conduct. And this point expresses nicely the difference between the inductivist and the Popperian. The former wants to establish his hypothesis:

> he hopes to make it *firmer* by his evidence e and he looks out for '*firmness*' – for '*confirmation*'. At best, he may realize that we must not be biased in our selection of e: that we must not ignore unfavourable cases; and that e must

comprise reports on our *total* observational knowledge, whether favourable or unfavourable.[21]

The Popperian goes in a different direction. He does not want to establish his hypothesis: he wishes to submit it to the most severe tests.

4.4   Corroboration revisited

In *RAS* Popper devotes an entire chapter to the problem of corroboration (determining whether there exists a measure of the severity of the tests that a theory has undergone, and showing that this measure does not satisfy the formal laws of the probability calculus). He points out that corroboration is one of the ramifications of the problem of induction. Although corroboration is not logically linked to the latter, many thinkers have brought in the issue guided by an 'inductive prejudice' and persuaded by a mistaken solution of this problem. According to this view, inductive logic is nothing more than probability logic, which, at its turn, hands over the long-awaited solution to Hume's problem. Probability logic is the logic of uncertain inference based on premises known with certainty. With the formula $P(h, e) = r$ we express the degree to which our certain knowledge of the evidence e rationally supports our belief in the hypothesis h.

This kind of inference is not uncommon. And its conclusions do go beyond what is asserted in the premises (for they do not fully entail the conclusions). Hence, one might solve Hume's problem by developing a theory of probability that allows assessing the probability of inductive inferences given the evidential support. In the formula above, this would amount to determining the precise value of r, assuming that it can take any fraction between 0 and 1, and that we represent good probability with the values greater than 0.5, gradually increasing the figure up to the maximum of 1. Popper offers two responses to the foregoing proposal. The more general one involves his criticism of the subjective interpretation of probability (the idea that probability expresses incomplete and uncertain knowledge), the other shows why the inductive theory of probability cannot work. I shall be concerned here exclusively with the latter.[22]

On Popper's view, the belief in probability logic springs from the mixture of two different ideas: the first supports our decision to distinguish hypotheses according to their performance in the testing ordeal; the second claims that there exists an inductive probability logic.[23] Plainly expressed, the first one says that some hypotheses are well tested by experience and others are not, that there are hypotheses that have been subjected to tests and have failed them, while there are some hypotheses that have not been tested so far, and that we can grade (with the appropriate marks) hypotheses regarding the tests passed. According to Popper, this idea is basically correct, and is defensible but not logically connected with the problem of induction. The second idea – the belief in inductive probability – is indefensible for logical and mathematical reasons. It should be rejected not only because it is false but also because it embodies a mistaken attitude towards science.

Let us suppose, for the sake of argument, that the formula $p(h,e)$ can express adequately the amount of support that evidence e provides for our hypothesis h. It is not difficult to show that we would obtain high probabilities only for hypotheses that give very little information about the natural world. For if h goes far beyond e, the probability will radically decrease until it becomes zero for every universal hypothesis. So, interesting hypotheses (i.e. truly informative universal hypotheses) will not fare well in the light of the calculus of probability. The alternative left to cope with this unwelcome result – advance hypotheses which do not go very far beyond evidence – illustrates what Popper has in mind when he says that devotees of probability logic are committed to a wrong picture of science.[24] The epistemologist who aims at hypotheses of high probability has to sacrifice the most inspiring characteristic of science: he has to renounce the ability to propose bold hypotheses and embrace instead the safe comfort of highly probable (but cautious) statements that advance greatly restrictive claims.

On the other hand, Popper recognizes that the words 'probable' and 'probability' are often used in a sense very close to what he proposes for his new expression 'degree of corroboration'. But there is a significant difference between the word 'probable' and the word 'corroborable' that explains why he decided to introduce the latter to designate an idea that cannot be conveyed by the former.[25] Consider the statement 'the probability of throwing 6 with a fair die is 1/6' (call it S). S does not inform us of anything about the severity of the tests that a hypothesis has passed (or failed); it gives us information about the *chances* that an event will occur. Of course, this type of statement satisfies the laws of the calculus of probability. In particular, it satisfies the axiom of monotony because its probability decreases with increasing logical content of the statement. But the probability of a hypothesis, in the sense of its degree of corroboration, does not satisfy the laws of the calculus of probability. For even believers in the logic of probability would not hesitate to say that a well-corroborated hypothesis is more 'probable' than a less well-corroborated competitor notwithstanding its greater logical content (i.e. its improbability) and in spite of the low chances that an event described by a highly testable hypothesis will occur.

In sum, the theory of corroboration solves the practical problem of induction ('when do we – tentatively – accept a theory?') by holding that we accept a theory when it has passed the *most severe* tests we have been able to design, and more especially when it has done it better than any of its competitors.[26] In Popper's opinion there is no need to carry this problem any further, and as suggested before, the subsidiary problem of whether it is possible to assign a precise measure to the severity of the tests reveals itself as unimportant. As a general rule, recall that the degree of corroboration of a hypothesis varies in direct proportion to its degree of testability. But testability can be measured by the content of the hypotheses (i.e. its absolute logical improbability); hence degree of corroboration is closely related to all and each of these alternative formulations of falsifiability. The upshot is that good epistemologists, appealing to the critical or falsificationist attitude, would always attempt to overthrow scientific conjectures: 'one looks for falsification, or for counter-instances. Only if the most conscientious search for counter-instances does not succeed may we speak of a corroboration of the theory.'[27]

A few words about the notion of 'background knowledge' are in order here. In the previous exposition of corroboration, we have referred exclusively to the relationship between a hypothesis h and the evidence e that might corroborate it, as though corroboration were merely a matter of logical compatibility between a particular hypothesis and the class of accepted basic statements. But there is something missing from this picture, because (as mentioned before) every observation whatsoever is mediated by a pre-existent, antecedent theory. So, when making decisions on the degree of corroboration of a hypothesis, we need to consider the role of a further variable (call it b) that stands for our background knowledge. Thus, we now say that e, corroborates h to degree x in the light of b. And for the latter we mean

> any knowledge (relevant to the situation) which we accept – perhaps only tentatively – while we are testing h. Thus b may, for example, include initial conditions. It is important to realize that b must be consistent with h; thus, should we, before considering and testing h, accept some theory h which, together with the rest of our background knowledge, is inconsistent with h, then we should have to exclude h from the background knowledge b.[28]

Although sometimes it might be difficult to motivate the decision to exclude some hypothesis from b, it is clear that one cannot test h assuming a set of initial conditions that is incompatible with it. On the other hand, by bringing b into the equation we can give a more satisfactory formulation of corroboration. This can be phrased in the following way: to attribute a high (positive) degree of corroboration to a hypothesis, we now demand that e follows from h in the light of b, without following from b alone, and that e be logically improbable (or that the probability – in the sense of the calculus of probability – of e given b be very low, if e is to offer strong support for h and is to account for the severity of the test that it has passed). In other words, this formulation requires that only evidence that is improbable in the light of our background knowledge would be accepted as significant to the effect of assigning a positive degree of corroboration to a hypothesis or to increase its degree of corroboration.

Although measures of corroboration do not satisfy the laws of the calculus of probability, Popper offers a definition of 'corroboration' that employs the relative probability of h, e, and b. Let '$C(h, e, b)$' stand for the degree of corroboration of h, by e, given b. In this definition we assume that e is a genuine test (or its result) of h, and we measure the severity of e in such a way that $C(h,e)$ increases with $Ct(e)$ (the content or absolute logical improbability of e). We also assume that e follows from hb and that $p(e, b) < \frac{1}{2}$. On the other hand, we assume that the support given by e to h becomes significant only when the probability of e given hb minus the probability of e given b is greater than $\frac{1}{2}$. Thus, the degree of corroboration (C) of h by e in the presence of background knowledge b can be expressed in the following equation:[29]

$$C(h, e, b) = \frac{p(e, hb) - p(e, b)}{p(e, hb) - p(eh, b) + p(e, b)}$$

Figure 4.1

The denominator in the definition is simply a normalization factor $(-1, 1)$ to obtain better the desired results, secure that only hypotheses and tests that are highly improbable receive good marks and remove what seems to be a fault: take a case in which e falsifies h and h may well be very improbable relative to b. Without normalizing, the measure becomes approximately zero. But this would be the value for a case in which e is a consequence of b. Introducing the normalization factor changes the measure of corroboration to -1 for the first case enabling us to make the appropriate distinctions. Popper finds the above formula satisfactory because it yields results that are compatible with the general lines of his theory. For example, when e supports h (given the background knowledge b) then $C(h, e, b)$ is positive. Moreover, when $p(e, hb) = 1$ and $p(e, b) = 0$, we obtain the maximum value for $C(h, e, b)$, namely 1 (under the complementary assumption that $p(h, b) = 0$). Contrariwise, if e is a tautology or a logical consequence of b, then e cannot corroborate or undermine h and the value of $C(h, e, b)$ becomes 0. The same result obtains if h is a tautology (since it would be neither corroborated nor undermined by e). By contrast, if e falsifies h in the presence of b, then $p(e, hb) = 0$; and if e reports the result of a severe test, then e would be very improbable relative to b, making $p(e, b) \approx 0$. In this case, and assuming that e is incompatible with hb, the degree of corroboration will be equal to -1.[30] Accepting these conventions, we are able to make sense of the expression 'positive degree of corroboration' and expect naturally the occurrence of its converse: 'negative degree of corroboration'. On Popper's view, the former should be assigned to well-corroborated hypotheses and should vary, in accordance with the severity of the tests, in the interval between 0 and 1 (including the rightmost extreme). In a way parallel to what we did to represent the degree of falsifiability, we can imagine a horizontal continuum that goes from -1 to $+1$ in the extremes with 0 in the midpoint and accommodates theories according to their degree of corroboration. We reserve the right extreme ($+1$) for the best-corroborated theories. The middle-point (zero) is reserved for tautologies, and the left extreme (-1) should be used for theories that were once corroborated and become falsified at a later moment. Popper thinks that his measure C shows both that the degree of corroboration of a theory is an evaluation of its performance in the empirical tests it has undergone, and that the degree of corroborability is equal to the degree of testability.[31] However, this assimilation is somewhat problematic, as I shall show below.

Popper defined the degree of falsifiability (Fsb) of an empirical hypothesis h as a proper fraction in the open interval between 0 and 1. The exclusion of both extremes was in order because he wanted to reserve the former limit for non-empirical statements (tautologies) and the latter for self-contradictory statements. Given the close logical relationship between testability and corroborability, one should expect that the same diagram suggested for Fsb would do an equally good job for the latter (after all we are dealing with concepts that stand in the *same* logical relation to theories). However, this is not the case. To see the reasons let us make some distinctions first. If 'degree of falsifiability' expresses the measure of a disposition (a potentiality) it does not make sense to apply this notion to theories of all kinds. More clearly, while distinguishing theories according to their degree of falsifiability might be significant when we are dealing with non-falsified theories and need to select between competitors, once the theories are refuted further distinctions (based on this logical feature) become irrelevant since all falsified theories (*qua* refuted) are equal.[32] A question arises here: where should we put falsified theories in the diagram for Fsb? (I am using $Fsb(e)$ for 'the degree of falsifiability of empirical theory e'):

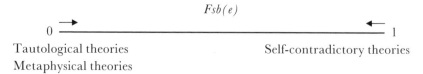

Figure 4.2

According to Figure 4.2, the degree of falsifiability of an empirical theory is always greater than 0 and less than 1. Thus, theories can be rank-ordered in the horizontal continuum according to their respective degree of falsifiability. Consider as a possible answer to the question just formulated to place falsified theories in the rightmost extreme. But this would amount to admitting that a falsified theory reaches the maximum degree of falsifiability (which was reserved for self-contradictory theories) as though (after falsification) it had become formally contradictory. Though this is *never* the case, the decision to treat a falsified theory on a par with a contradictory one can be motivated by means of an analogy: for any falsified theory it is possible to point out a contradiction between h and e (the falsifying hypothesis) in the light of b. On the other hand, since it is always possible to single out the contradictory pair of statements in these theories, we might be able to rescue them by eliminating the contradiction, for the theories might still be testable with regard to some other part of their empirical content.[33] Of course, neither can we put the falsified theory in the leftmost extreme, nor can we put it in any place between 0 and 1. So, it seems that this should be the right answer. However, there is something awkward in the locution 'falsified theories have the maximum degree of falsifiability' (which is how we can read the decision of treating these theories on a par with self-contradictory statements). It is clear that this could not be the case if 'maximum degree of falsifiability' means 'ease of falsification', as we may suppose from Popper's original definition and his stated motive to restrict the attribution of that measure to self-contradictions. What to do, then, with these sorts of theories? Perhaps we should recall the sharp distinction between falsifiability and falsification that Popper made before, and answer the question in a different way: falsified theories cannot be properly represented in the diagram for *Fsb*, because the concept does not apply to them (they cannot be assigned any degree of *Fsb* in a significant way). They *had* a high degree of *Fsb*, in the past. Once they are refuted, we do not need to call them other names than 'falsified'. I suspect that a failure to make a similar distinction lurks behind Popper's suggestion about how to draw the corresponding diagrams for degrees of corroborability (*Crb*). To summarize: one cannot represent adequately the parallel between *Fsb* and *Crb* with the diagram of Figure 4.2 because the maximum degree of *Crb* corresponds to a theory which is highly falsifiable (though not contradictory) but which can survive testing, and Figure 4.2 has no space to accommodate the last sort of information. Since Popper frequently switches between *Crb* and corroboration (*Cr*) in this debate, let us consider the latter.

We can see that neither would Figure 4.2 work to represent degrees of *Cr* by considering that a well-corroborated theory is nothing more than a highly falsifiable theory that has passed a severe test. Obviously this sort of theory cannot be placed in any of the extreme points (let us put aside, for the moment, the interval between 0 and 1). In his discussion of corroboration, while still maintaining the parallel between corroborability and falsifiability, Popper suggests a diagram such as Figure 4.3 below, to rank theories according to their respective degree of corroboration (I let the abbreviation $Cr(e)$ stand for 'degree of corroboration' of e):

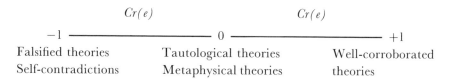

Figure 4.3

Take the case of corroborated theories. In so far as we accept h as a well-corroborated theory, agree in the severity of the undergone test and assume that h is still falsifiable, we can ascribe to it the maximum mark for corroboration. Tautological (as well as metaphysical) theories are not corroborable, so we ascribe to them 0. Given that corroborability and falsifiability increase and decrease together we need to make a decision concerning those theories which are falsifiable in the (absolute) highest degree. Popper puts them in the leftmost extreme (symmetric to the mark used for well-corroborated theories), treating falsified theories on a par with self-contradictions under the assumption that they fail every test to which they are submitted. Disregarding some complexities, the same place can be used for falsified theories that passed really severe tests before falsification. I have placed other theories in the interval between $(-1, 0)$ and $(0, +1)$ to indicate that we may rank-order theories according either to their positive or their negative degree of corroboration, when these degrees are different from the extreme cases. However, I do not find Popper's suggestion (as represented in Figure 4.3) satisfactory for two reasons: firstly, because it seems to conflate corroborability with corroboration; secondly, because it does not capture correctly the logical relationship between falsifiability and corroborability.

If we distinguish carefully between corroboration (Cr) and corroborability (Crb) (as we did before with falsifiability and falsification), and we introduce a slight change to Popper's suggestion, I think we can solve the problem and capture better the nature of the concepts we are dealing with. The change I want to propose consists in using a diagram as close as possible to Figure 4.2, ignoring the recommendation to work with negative degrees. Accordingly, we can rank-order theories by their respective degree of corroborability as follows:

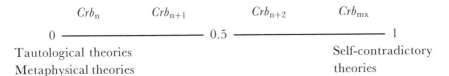

Figure 4.4

Under this convention, tautological as well as metaphysical theories are accorded a degree of corroborability equal to 0, since they are not in general corroborable. Self-contradictory theories are accorded the maximum degree of corroborability, only when we consider the relation between Crb and Fsb, since they are highly (and easily) testable. For other empirical theories we use the interval, making the degree as big as needed (indicating the maximum with the subscript mx) without equating it to 1, for the same reasons we did not want to equate the highest degree of Fsb of a consistent empirical theory to 1. We can see that Figure 4.4 captures the logical relationship between Crb and Fsb, and it seems to fit our intuitions. To represent degrees of *corroboration*, we need a different diagram.

To begin with, note that a corroborated theory should have survived a test whose severity we want to reflect in our ascription of the corresponding degree of corroboration (Cr). Recall, also, that corroboration by itself is not sufficient to support our choosing a theory from a group of competing theories. We want both, good degree of corroboration and Fsb.[34] But a corroborated theory might be wanting with regard to its informative content, or it might be such that it no longer has a good degree of Fsb; or it may have been refuted on another count after some spectacular successes, and so on and so forth. Given that Cr tells how well the theory has passed the tests and how severe they were, we may see that the following conventions are reasonable. Tautologies are non-corroborable, so they can be accorded a zero degree of Cr. Self-contradictions were highly Crb solely because of the ease of testing, but they never pass any test, and the tests that they *fail* are in no way severe, so I propose to exclude them from Figure 4.5 below. The rest of the theories can be ranked in the interval $(0, 1)$ (inclusive of 1), where the subscripted index indicates increasing degree of Cr (I am using 'Cr-w' for 'well corroborated'):

```
        Cr_n       Cr_{n+1}       Cr_{n+2}    Cr_{n+3}    Cr-w
         ft          ft            Fsb         Fsb         Fsb

0 ─────────────────────────── 0.5 ─────────────────────── 1
Tautological theories
Metaphysical theories
```

Figure 4.5

In my opinion, Figure 4.5 expresses neatly what one should infer about Cr, being respectful of the difference between it and Crb. We assign 1 as the maximum Cr to a theory under the assumption that in this case Crb equals content (instead of being merely proportional to it). To see how Figure 4.5 captures better the notion of Cr, we only need to add the following caveats: (1) any assignment of Cr is temporal and should be *indexed* to the particular test passed; (2) we assign the maximum value of Cr only to theories that have passed a severe test and are still falsifiable in a significant degree; (3) falsified theories (ft) that passed some tests are ranked according to the severity of those tests, receiving a mark less than 0.5 in all cases; (4) theories are ranked (and compared) only with respect to the same sort of tests; and (5) theories can retain or increase their Cr (within the restrictions of (2) and (3) as a result of their performance in additional but different severe tests to which their competitors have been submitted also.[35]

Neither the diagrams above nor the one in Chapter 2 appear in Popper's writings, but they are drawn according to his suggestions or to defensible revisions of certain logical relations holding between the respective definitions of Fsb and Crb. In my opinion, the blemishes in Figure 4.3 above are due to Popper's interest in relating Fsb and Cr. He wants to establish a direct connection between Fsb and Cr that can block the popular objection that one cannot satisfy both desiderata simultaneously. Instead he gives us a relation between Fsb and Crb (a relation that is easier to get) and by switching back and forth between Crb and Cr (perhaps inadvertently) his prose becomes extremely obscure and he can hardly make the point. Now, one cannot deny that there is a relation between Fsb and Cr, but is not the class of relation that Popper requires for his point. For only one reason: degree of Cr is a function of the severity of the test survived, whereas high degree of Fsb is a function of the cardinality of the class of PF plus the easiness of testing. If this is the case – and it should be, according to the definitions – when a highly Fsb theory survives a severe test it becomes well corroborated – with respect to *that* particular test – but we need to show independently that it is still Fsb and that cannot be done with respect to the test passed.

On the other hand, when spelling out the notions of Fsb and Cr the use of a modal or an actual context makes an important difference in one's theory of science. To exemplify the growth of empirical knowledge a Popperian wants Fsb (modal) combined with Cr (actual). From this point of view, it is easy to distinguish his position from the stance of the inductivist. The latter does not value high falsifiability because he is looking for verification. But, although valuing verifiability, the inductivist would not be satisfied with mere corroborability either; he does not want a theory that can possibly survive a severe test. He wants a theory that can possibly be verified in experience, and this takes us to a nice contrast between falsificationists and verificationists: as explained above, the degree of Cr is directly related to the nature of the tests survived. Those who claim that corroboration is just another name for induction have overlooked the fact that the degree of Cr of a theory is determined primarily by the severity of the tests that it survives, and only in some special cases by the number (recall that the cardinality of the tests increases corroboration very little). By contrast, verificationists – in so far as they subscribe to induction – are committed to saying that the number of instances of verification increases the degree of 'corroboration'.[36]

4.5 Criticisms of Popper's notion of corroboration

The commonest criticism simply says that the notion of corroboration expresses the statistical (inductive) probability of a given hypothesis in light of evidence.[37] We have already seen how Popper disposes of this criticism and why he sees it as an indirect consequence of the unwillingness of some critics to accept his solution of the problem of induction. There are some authors who claim that given two hypotheses, neither of which is falsified by the evidence, Popper's notion of corroboration regards the more falsifiable as the better confirmed.[38] But this assumes the false premise that being 'more falsifiable' entails being 'better corroborated' and it obviously fails to distinguish between corroborability and corroboration. What Popper says is that a higher degree of Fsb entails a better chance that testing will result in a positive degree of corroboration, and more importantly, that although a high degree of testability determines how severely we can test a hypothesis, it does not determine how successfully the hypothesis may withstand the severest possible tests. On the other hand, many of the best-articulated criticisms of corroboration take one of two lines: (1) they point out that once a theory has been well-corroborated it does not seem to satisfy the *desideratum* expressed by the equation 'corroborability = testability = empirical content'. For if the degree of testability of h is a function of how unlikely it is that h survives a testing situation, when this occurs (i.e. when h becomes well-corroborated) we lose grounds to attribute to h both high degree of Fsb and a good mark in Cr, because we know now that h passed the test (i.e. h is no longer falsifiable with respect to *that* test). We can put forward a similar worry by saying that well-corroborated hypotheses (*qua* corroborated) are not, in general, highly falsifiable. (2) They claim that Popper cannot have corroboration without inductive assumptions. As I shall show shortly, critics argue that one cannot assess the performance of a hypothesis in a testing situation ignoring unstated expectations about the results of repeated tests, not to mention the more general assumptions about the regular behaviour of nature. In this way – the objectors charge – Popper sneaks induction back into his theory. Let me discuss these criticisms appealing to the works of some influential thinkers. I shall start with Putnam's objections.

4.5.1 Putnam: theory is not an end in itself

In Putnam's opinion, the scientific methodology that Popper advocates is both partly right and partly different from common (inductivist) philosophy of science. However, in what Popperian methodology is right about does not differ from the common view, and about what it differs from the common view it is mistaken. Popper, like the inductivists, accepts that scientists put forward general laws and that they test them against observational data (and here Popper is right). As we have seen in previous chapters, the main difference (with respect to common philosophy of science) can be summarized by saying that Popper only admits falsification, so that Hume's problem does not arise for Popperians, for they do not intend to verify laws but only to falsify them (and here – Putnam contends – Popper is wrong). Further, Popper claims that when a scientist corroborates a law he is not entitled to assert that it is true or even probable. And this is because an appraisal of corroboration is just a report on how well the corroborated statement has withstood a test. In addition, Putnam believes that Popper neglects the pragmatic side of science because in his scarce references to this issue he reduces the application of scientific theories to another test of the laws.[39] We can see the inadequacy or Popper's philosophy of science – Putnam maintains – by noting that it makes science practically unimportant and incapable of producing understanding of the world. The former, because scientists would never tell us whether is it safe to rely upon a theory for practical purposes. The latter, because limiting our judgement to cautiously ascertaining that a provisional conjecture has not been refuted yet does not amount to understanding anything. We can avoid both mistakes, however, by recognizing that theory cannot be an end in itself, admitting that science aims to produce projectable – i.e. useful – knowledge about the world, and accepting that the distinction between knowledge and conjecture has important consequences in our lives.

Certain features of Popper's theory of science bother Putnam. I shall focus here on two problems. On the one hand, Putnam finds Popper's insistence on the non-verificationist character of his notion of corroboration quite perplexing. On the other, Putnam thinks that even if we could make sense of Popper's notion of corroboration, his epistemology would fail because scientific theories in such an epistemology do not imply predictions at all. I shall discuss these problems in the same order I introduced them.

According to Putnam, Popper's notion of corroboration presupposes induction. And this is so despite Popper's insistence to the contrary. For we can hardly deny that people apply scientific laws in many situations, let alone that such applications involve the inductive anticipation of future success. In addition, reports on corroboration are barely distinguishable from reports on verification, because they depend on the rule: '*look at the predictions that a theory implies, see if those predictions are true*', which is common to both the deductivist and the inductivist schemas.[40] Of course, Popper gives us a story about how induction cannot support scientific laws and advises us to regard scientific statements as tentative guesses. But the story seems implausible while the advice is neither reasonable nor does it represent what scientists do. To see this clearly, recall that in Popper's epistemology, when scientists accept a theory they select the best-tested theory provided it is also the most (logically) improbable. In Putnam's opinion this constitutes a weird use of the term 'improbable' and overlooks the fact that one among all the theories implying an improbable prediction could be the most probable. More importantly, in the scenario in which several theories have survived a particularly severe test, it is not clear how to choose the most improbable theory *after* the prediction turns out to be true.

Since corroboration (as well as verification) requires a deductive link between a theory and its predictions, one way to discredit either would be to show that, in general, theories do not make predictions.[41] Putnam intends to do this by using as an example the theory of universal gravitation (UG). On his view, UG alone does not imply a single basic statement because it says nothing about what forces other than gravitation may be present in any system of bodies related in such a way that each body a exerts on every body b a force F_{ab}, not to mention that these forces are not directly measurable. If we want to apply UG to an astronomical situation (e.g. to deduce the orbit of the earth), we need to make some auxiliary assumptions that simplify the theory stipulating its boundary conditions:

(i) No bodies exist except the sun and the earth.
(ii) The sun and the earth exist in a hard vacuum.
(iii) The sun and the earth are subject to no forces except mutually induced gravitational forces.[42]

Only by using the conjunction between UG and the auxiliary statements (AS) does it becomes possible to derive certain predictions (e.g. Kepler's laws), and by making i–iii more realistic the predictions can be improved. Putnam urges the point that AS are not part of the theory and indicates that usually they are false statements (like i–iii above) which are open for revision. He sees the well-known story of the discovery of Neptune as a confirmation of the fact that when a theory appears to be in trouble, scientists modify the AS rather than attempting a revision of the theory, for when a theory has been extremely successful, a few apparent anomalies furnish no reason to reject it. So, keeping the theory and modifying the AS is common (and good) scientific practice, unless there is a better alternative theory available. If the picture just described is accurate, it will affect Popper's doctrine and show that

> The law of Universal Gravitation is *not* strongly falsifiable at all; yet it is surely a paradigm of a scientific theory. Scientists for over two hundred years did not derive predictions from U.G. in order to *falsify* U.G.; they derived predictions from U.G. in order to *explain* various astronomical facts. If a fact proved recalcitrant to this sort of explanation it was put aside as an anomaly (the case of Mercury). Popper's doctrine gives a correct account of neither the nature of the scientific theory nor of the practice of the scientific community in this case.[43]

Claiming that Popper does not intend to describe scientific practice but to say what scientists should do (by way of proposing methodological rules) does not block the above objection since there are no compelling reasons (in the example of UG) that make the abandonment of the theory the right move. Given that scientists could not attempt a falsification of UG due to practical and epistemic limitations, and that the theory predicted many other orbits correctly, it would have been a mistake to reject it because of a deviance in the orbit of Uranus. Moreover, since it is possible to overlook the role of non-gravitational forces or fail to consider all the gravitational forces that are relevant to an astronomical situation, we might be more justified when admitting that astronomical data support UG than we might be regarding its falsification. On the other hand, failure to pass an experimental test does not prove that a theory is incorrect (theories are not strongly falsifiable). With regard to the contrast between falsification and verification, Putnam concludes that the former is not more conclusive than the latter. As we can infer from this argument, Putnam sides with Kuhn when advancing the claims that scientists rarely, if ever, try to falsify a theory and that anomalies do not afford strong reasons leading to the rejection of an empirical theory.[44]

According to Putnam, one problem for both deductivist and inductivist philosophies of science is that they fixate on the situation in which we derive predictions from a theory and test them in order to falsify or verify the theory. But science seems to follow a different pattern, one in which we take the theory and the fact to be explained as fixed and look for facts about the particular system that will afford us the explanation of the fixed fact on the basis of the theory. Although he does not produce any argument to this effect, Putnam thinks that we cannot measure falsifiability in terms of logical improbability or content, as Popper claims. Furthermore, Putnam admits that UG has probability zero on any standard metric (as Popper attributes to universal hypotheses), while he denies that the theory rules out any basic sentences. Regarding the selection of theories, Putnam believes that scientists only choose 'the most improbable of the surviving hypotheses' in the trivial sense that all universal laws have probability zero, whereas this is never the case in any measure of probability. Thus, the Popperian sense that attributes low (logical) probability to falsifiable statements does not make sense if that concept means probability assigning equal weight to logically possible worlds.[45] Finally, Putnam agrees that good theories make their way to science only if they have substantial, non-*ad-hoc* explanatory successes. However – he proceeds – given that the primary importance of theories is to guide practice, and that we determine the correctness of ideas in that realm, success in general accords better with inductivist accounts of science because they 'stress *support* rather than *falsification*'.[46]

In his 'Replies to my Critics' Popper explains how Putnam goes wrong. In particular, he is concerned with Putnam's starting-point that nothing can be deduced from isolated theories, so that if we are interested in predictions we need to devise some auxiliary statements that are to be conjoined with the theory. Without further qualifications, we can see that this is not a fair criticism of Popper, since he wrote in *LScD* that nothing like an observation statement follows from a universal law alone, without additional initial conditions.[47] Moreover, he explicitly equates 'initial conditions' to 'auxiliary statements' and avows that positive 'instances' of a law cannot be regarded as corroborating or confirming it because they can be had for the asking.[48] Conversely, under certain circumstances theories (if they contain existential statements) can entail empirically verifiable statements, or (if they are strictly universal) can entail the negations of basic sentences (which in general are not basic sentences themselves). By attentively considering the latter, on the other hand, we can show that Putnam is not very careful when claiming that nothing whatsoever can be deduced from UG unless we adopt auxiliary statements, for it is possible, indeed, to deduce the negative statement 'There does not exist any pair of bodies x and y that do not mutually exert a force F_{xy}.' We can say something along these lines regarding Putnam's conclusion that no theory is strongly falsifiable. Actually, when he makes the conjecture that scientists would have abandoned UG if the world had started to act in a markedly non-Newtonian way, he seems to disprove his own claim, for all that Popper needs for falsifiability is the incompatibility of the theory with a basic statement (that describes an event) like the one Putnam just identified.

Putnam contends that there is an inductive quaver in Popper's description of corroboration and he suggests that any tentative acceptance of the result of an investigation can be called 'inductive'. Let us put aside, momentarily, the former charge. The latter betrays a misunderstanding of Popper's description of what takes place when a scientist selects a theory. According to Popper, choosing a theory from a set of competing alternatives 'is an act of preference, and any preference of a theory over another is in its turn a conjecture (of a higher order)'.[49] So, one may choose a theory without being committed to taking the theory as inductively established, and calling the theory 'best' only means that it is both the best-corroborated and the best-testable theory left. Before turning to my own assessment of Putnam's criticisms, let me report also that Popper points out in his reply that Putnam's assertion 'Kepler's laws are deducible from UG plus the AS' is historically false, for those laws are not only incompatible with UG but were corrected by the theory. To Putnam's accusation that when confronting falsification one can never determine whether to put the blame on the theory or on one of the AS, so one is not sure which has been falsified, Popper replies that: (1) we should regard the whole theory as falsified; (2) determining whether the culprit of falsification is the theory or one of the AS is also a conjecture; and (3) replacing or modifying the AS might afford us a way of substituting the theory for a better alternative, but that configures an entirely different situation.

I find these replies both right and clear. However, they can be reinforced with the following points. To begin with, Putnam takes the assumption that a well-corroborated theory may still remain highly testable after corroboration as utterly odd. He wonders how it is possible to select the most improbable of the *surviving* theories *after* the prediction has turned out to be true. The answer to this quandary is not difficult. We do not consider a theory logically improbable because of the tests it withstood (which make it corroborated). A theory is logically improbable due to its non-tested (negative) predictions, and it is to those that we need to turn our attention when attempting to select the most improbable of the surviving theories. Of course, if Putnam has in mind a well-corroborated theory that has no severe tests left, then his worry would be justified. However, although Popper has acknowledged that the degree of corroboration of a well-tested theory grows stale after some time, we should not expect that kind of situation for genuinely universal theories with good explanatory power and high empirical content, since they should have plenty of prohibitions to keep the tests running for a good deal of time. On the other hand, given that in Popper's sense empirical science does not aim at establishing definite truths,[50] we must be prepared to see even the most successful theories deteriorate and be replaced by better competitors, and surely decreasing falsifiability qualifies as one sign of deterioration.

I do not think that Putnam's criticisms establish that one cannot have corroboration unless one makes some inductive assumptions. Putnam seems to think both that science is meaningful only if aimed to practical matters (something he deems extraneous to Popper's project) and that asserting that a theory has passed a test commits the evaluator to expect that it will pass additional future tests. The first belief seems to me inadequate, but even if it were adequate it would not represent faithfully Popper's view. There is as much room for practical success in Popper's theory of science as one could want. The second belief is simply wrong. Popper has repeatedly claimed that his notion of corroboration does not involve any prediction, whatsoever, about the future performance of the theory on further tests. Moreover, if we consider the basic principles of the theory of corroboration, we could see that no presupposition is made (or needed) about how a theory will perform on future tests. Since appraisals of corroboration only report the severity of the tests and how well the theory stood up to them at a given moment of time, any inference about future performance is neither required nor warranted. Now, if Putnam is complaining because people approach those appraisals inductively, then he is barking up the wrong tree. Popper gives the definitions and advises on their appropriate application; he cannot control, however, how one uses them any more than makers of any device can guarantee that consumers will use their products in the intended way. In fact, careful reflection upon Popper's theory of corroboration reveals that, when we are concerned with questions pertaining to the status of theories, we can explain all our decisions by appealing basically to deductive procedures. Finally, I need to remind those who mention the issue of the belief (or expectation) in the regularity of nature as an instance of an inductive assumption implicit in the theory of corroboration, that Popper treats this belief as a metaphysical rule. Furthermore, since he makes no claim for the generalizability of appraisals of corroboration, the objection fails.

4.5.2   Lakatos: corroboration is a measure of verisimilitude

We can find Lakatos's criticisms of Popper's notion of corroboration in his long paper 'Falsification and the methodology of research programmes', which is restated with minor changes in his contribution to Schilpp's (1974) volume. In the latter, besides reiterating his general criticism of dogmatic falsificationism, Lakatos links corroboration to verisimilitude and presents a case against both notions. His elaborated story-like argument runs like this: earlier (inductivist) criteria of demarcation laid down the aim of science as the discovery of the blueprint of the universe. They took each partial 'discovery' as a step towards the goal and defined progress accordingly. When they realized the difficulties of obtaining the true blueprint (derived in part from the problem of induction), they settled for a more modest goal in terms of probability. By contrast, Popper does not specify the aim of science. He considers the game as autonomous and thinks that one cannot and need not prove that the game actually progresses towards its unstated aim. For Popper, it suffices to hope that the game will make progress and that we can pursue the game simply for its own sake.[51] The main asset of science was its capacity to detect error, and scientists were victorious in rejecting falsified theories and provisionally accepting those that were corroborated. But Lakatos thinks that the only function of high corroboration is to challenge the scientist to make every possible effort to overthrow a theory. As a result, Popper's theory of science offers methodological rules for either rejecting or accepting theories, and in so doing dissolves the problem of induction (without providing a positive solution).

According to Lakatos, it is clear that within the framework of his *Logik der Forschung* Popper cannot answer the question of what one can learn in the game of science.[52] Popper's motto that one learns through one's mistakes does not do it, because one needs a theory of truth in order to detect error and a way to establish whether theories increase or decrease in truth-content in order to determine progress. And Popper did not have at his disposal the logical machinery required to treat both problems by the time he published his main work. It was not until the early 1960s that he actively incorporated Tarski's theory of objective truth into his own philosophy of science, and that changed his stance drastically. By exploiting his theory of verisimilitude, Popper became able to state the aim of science in terms of truthlikeness, re-establishing the link between the game of science and the growth of knowledge. Besides he could overcome scepticism, avoid irrationalism and advance a more optimistic epistemology while providing, finally, a positive solution to the problem of induction: now we know that increasing degree of corroboration is the signpost of increasing verisimilitude. The problem, however, is that anyone who wishes to address satisfactorily the question 'What can we learn in the game of science?' appealing to methodological falsificationism, needs to invoke some principle of induction. For we can easily recognize progress

> by an inductive principle which connects realist metaphysics with methodological appraisals, verisimilitude with corroboration, which reinterprets the rules of the 'scientific game' as a – conjectural – theory about the signs of the *growth of knowledge*, that is, about the signs of *growing verisimilitude of our scientific theories*.[53]

Lakatos rephrases what he considers is Popper's positive solution to the problem of induction by pointing out that we can deem the works of the greatest scientists as exemplifying ways to increase our knowledge of the world (in terms of approaching Truth). Now, in Lakatos's opinion, Popper's methodological appraisals contain the hidden inductive assumption that 'if one lives up to them, one has a better chance to get nearer to the Truth than otherwise'.[54] In addition, one should interpret a high degree of corroboration as a sign that the scientists might be getting close to the truth, just as Columbus's sailors interpreted the birds above the ship as a sign that they might have been approaching land. Thus, inductive assumptions are present everywhere in

Popper's theory. Lakatos thinks that only an inductive principle can guarantee the stability of theory appraisals so that we can expect that attributions of falsification or corroboration remain, in general, unchanged through time. On the other hand, Lakatos thinks that after his 'Tarskian turn' Popper can correlate methodological appraisals (which are analytic) with genuine epistemological appraisals (which receive a synthetic interpretation) and claim that his positive appraisals can be seen as a conjectural sign of the growth of conjectural knowledge. But, according to Lakatos, Popper is reluctant to make this move; he does not want to admit that 'corroboration is a synthetic – albeit conjectural – measure of verisimilitude. He still emphasizes that "science often errs and that pseudoscience may happen to stumble on the truth".'[55]

Lakatos thinks that reluctance to accept that positive appraisals indicate (conjecturally) growing knowledge, commits a Popperian to a sort of scepticism under which one accepts that knowledge can grow, but without one's knowing it. The preference of theories to which the appraisals naturally lead, is also based on the inductive principle which asserts that science is superior to pseudo-science, but (regardless of this principle) Popper would classify a plausible statement like 'physics has higher verisimilitude than astrology' as 'animal belief'. To sum up, Lakatos presses three main charges: (1) he accuses Popper of relying on hidden inductive principles; (2) he speculates that had Popper been acquainted earlier with Tarski's theory of truth, he would have given a positive solution to the problem of induction since the time of his *Logik der Forschung*; and (3) he maintains that the correlation between verisimilitude and corroboration (which he treats as a measure of verisimilitude) provides foundation for an inductive (or quasi-inductive) principle that asserts the epistemological superiority of science over pseudo-science.

I find the three charges unwarranted. Let me spell out some reasons for this contention. Given that corroboration appraisals can be explained in deductive terms (inasmuch as they can be derived from a given hypothesis and the accepted basic statements) and that they merely report on the way a theory has passed a particular test, I fail to see where the inductive assumptions are supposed to be. Lakatos makes a big deal of the *inferences* that can be supported by ascriptions of corroboration and urges the point that in order to select a theory one needs to use induction but, quite frankly, and considering all we have discussed on corroboration in this chapter, I do not see that that is the case. If the appraisals (which, strictly speaking are not 'inferences') are used to advance tentative conjectures (as the preferences are) about the theory's potential we can keep the game within the constraints of deductivism. Furthermore, making inferences about the reliability of a scientific theory may require inductive assumptions, but we need to recall that Popper explicitly rules out the view that theories can be considered as reliable, likely or probable. The second charge is a bit more complex because it deals with the evolution of Popper's ideas and sustains different claims about the possibility of providing a solution to the problem of induction once this has been separated from the problem of demarcation. I shall say that, in general, Lakatos's speculation is not supported by textual evidence. We can read in many places that Popper learned about Tarski's theory of truth as early as 1935. Moreover, he referred to this theory in his Bedford College lectures of the same year, and in print in his *Open Society* (published in 1945), so the claim that Popper only incorporated Tarski's theory of truth into his own philosophy of science around the late 1950s or the early 1960s is, to say the least, disputable.[56] However, this could be considered merely a minor point. What I find more troubling is the denial that Popper can give a (positive) solution to the problem of induction without his theory of verisimilitude and incidentally without appealing to Tarski's theory of truth. If all that Lakatos is claiming is that Popper does not solve the problem of induction but only dissolves it, then I think I can agree with him. But, setting aside the issue of whether Popper has actually given a solution to such a problem, one needs to acknowledge that just how what Popper wrote about this topic in *LScD* can be reformulated so that it relies explicitly on the notion of truth; what Lakatos calls 'the positive solution' can equally be reformulated with no appeal to the notion of truth. To make this point clearer, let us remember that, according to Popper, Tarski's theory of truth enriched his understanding of certain logical problems but did not modify his epistemology.[57] The third charge deserves some careful consideration.

Lakatos finds it odd that Popper is willing to grant scientific status to an 'absurd' statement such as 'nothing can travel faster than the velocity of light', but treats a 'plausible' statement such as 'physics has higher verisimilitude than astrology' as 'animal belief'. This oddness could be avoided – Lakatos thinks – by connecting scientific standards with verisimilitude and taking into account that corroboration is a measure of verisimilitude (interpreting the word 'verisimilitude' in the regular sense of Popper's theory). But Lakatos's position seems to be the result of confusion of many different aspects. To begin with, Lakatos's plausible statement is neither a scientific statement nor a good candidate for such. It makes a meta-assertion about theories. Of course, nothing prevents us from using the guidelines of Popper's epistemology to analyse statements like that, but proceeding in this way would require taking the appropriate precautions and making the necessary distinctions, and the outcome would probably be quite different from what Lakatos's argument requires. Secondly, one usually attempts to conduct theory comparison with genuine rivals, and there are better choices than astrology to make the contrast Lakatos is suggesting. Thirdly, it is doubtful that Popper would consider a verdict about the status of astrology as mere animal belief, unless we define conjectural statements in those terms. But if we stick to what Popper has written on that topic, conjectures are the starting-point of the (rational and critic) method of science and afford us ways of improving our knowledge by subjecting them to careful criticism. However, the biggest problem for Lakatos's view of corroboration is how he understands its link to verisimilitude. In Chapter 10 of *CR*, Popper suggests that degree of corroboration *may* be taken as an *indication* of verisimilitude.[58] From that we can conjecture that the best-corroborated theory is also the one that is closer to the truth, and we may risk the same conjecture about unfalsified or falsified theories. This gives good grounds to deny that corroboration can be a measure of verisimilitude. Nonetheless, we can make such denial more forcefully. We only need to remember that degrees of corroborability are dependent on particular tests (which may or may not be performed) so that we accord a good degree of corroboration to a theory when it has passed a severe test. By contrast, verisimilitude depends on truth content, and although establishing that a theory is well corroborated can support our conjectures on verisimilitude, knowing that a theory has *passed* a particular test is not sufficient to substantiate an ascription of verisimilitude: we also need to have some information about the sets of true and false statements that follow from a given theory in order to do the latter.

4.5.3 *Derksen on corroboration*

As explained before, Derksen reports that a fair interpretation of Popper's theory requires the amendment of Claim I (we learn from our mistakes) with an enhanced Claim I*: 'only from falsifications, occasionally preceded by corroborations, can we learn'. Presumably, corroboration secures the empirical character of theories, but Derksen thinks that corroboration is problematic on the following three counts: (1) it gives rise to a question-begging argument; (2) it forces the Popperian to presuppose the form of induction exemplified by the belief that the future will be like the past; and (3) it cannot be appropriately combined with falsifiability. Let me spell these charges out.

In Derksen's reconstruction of Popper's theory, saying that a theory is the best corroborated (the most severely tested so far) amounts to making an appraisal in the light of our critical discussion. Singling out this theory (e.g. to select it) is only a measure of the rationality of accepting, tentatively, a conjecture. The problem is that the argument depends on the correctness of Popper's conception of rationality, and, hence, relies on Claim I* which assumes that corroboration achieves something. In a parallel way, we can see that scientific growth is possible only if genuine attempts to refute a theory fail occasionally (since corroboration is nothing but failed falsification), but this again presupposes that corroboration achieves something. I do not think it is difficult to reply to this criticism. Assume that what Derksen calls Claim I* captures adequately the idea that learning occurs through the replacement of false theories by corroborated theories. Given the provisional character of all conjectures (including those about the status of theories) one cannot say that Claim I* begs the question, because it only relies on the report of the testing results without requiring anything about what corroboration achieves (strictly speaking, Popper makes no claim whatsoever as to what corroboration may or may not achieve). Concerning Derksen's remarks on the correctness of Popper's theory of rationality, I think that this theory is not in question here. However, faced with this difficulty, I would choose Popper's own view: is there something more rational that the willingness to withdraw a claim if it does not sustain criticism? If Derksen has either a counter-example to this or a better story, I would love to know it.

Derksen's second charge is one of the variants of the more general accusation of the inductive dependence of corroboration. He claims that Popper's 'probabilistic' argument for the conjectural verisimilitude of well-corroborated theories can be non inductive about the past and the present but cannot be such about the future.[59] Let us briefly recast the argument. Popper says that it is highly improbable that a theory withstands a series of very varied and risky tests due to an accident; hence, it is highly improbable that the theory is very far from the truth. This judgement supports the guess that the theory *may* be closer to the truth (than some considered rival). But I do not see where the inductive assumption about the future lies. Popper's argument neither says nor assumes anything about the future performance of this theory in similar or different tests. Of course, one might read in the argument that if the theory has fared well in a series of tests, subjecting it to another test in the same series would yield similar results, and this can be enough to substantiate Derksen's accusation. Three observations are in place here, however: (1) there is nothing against the possibility of the theory failing the next test within the same family; (2) passing another test in the same family does not increase the degree of corroboration and should not affect our evaluation of the theory (in other words, inductively increasing the amount of favourable evidence for a test has no bearing on corroboration); and (3) Popper has explained tirelessly that corroboration does not provide information about the future. It only tells us about the performance of the theory in a particular test. So, Derksen's criticism does not follow.

The third charge reiterates a worry that I have already discussed in section 2.3. Roughly stated, it says that Popperians can never find a theory that satisfies the joint requirement of good corroboration *cum* high falsifiability, since the latter is bound to diminish with every significant increase of the former. If this is the case, either one lacks good reasons to choose between competing theories, or one needs to give up corroboration and use falsifiability as the only criterion of selection. In Derksen's words, since we cannot demand guaranteed success, we need to learn what is the strategy that offers us the best chance of making progress, and this may coincide with 'the best one available. It is disturbing, however, that the strategy proposed neglects – at least may neglect – the best corroborated theory, the one which so far survived the wildest tide of risky trials.'[60] I do not want to repeat my earlier arguments here. Let me just point out, that claiming that falsifiability is bound to diminish after every test passed does not do justice to Popper's theory. The assertion is true only when we consider a *particular* test, but a genuinely universal theory may remain highly falsifiable (in other respects) even after having survived a few severe (different) tests. In addition, any selection of theories should consider both degree of corroboration and falsifiability. Although it may not be possible to satisfy always the ideal case, we can look for theories that have both a good degree of corroboration and high informative content. If they are not available because all we have left are well-corroborated theories with low degree of falsifiability, it simply means that it is time to look for suitable alternatives somewhere else.

Notes

1. Before proceeding, let me make two terminological remarks. The first concerns the very term 'corroboration'. As Popper himself explains it in a footnote of *LScD* (p. 251) and reiterates it in *RAS* (p. 228), he used sometimes the term 'confirmation' to refer to the notion we are now discussing. The word seemed to be appropriate to convey the idea that Popper had in mind (what to say of a theory that has passed a severe test) and it was quickly accepted in the literature. But it was not neutral enough to do the job, and it soon showed strong and undesirable verificationist and justificationist associations. On the face of these problems, Popper replaced 'confirmation' by 'corroboration' and has used the latter exclusively ever since 1955. The second remark, which will be developed in due time, is related to the dissociation between 'corroboration' and 'probability'. It must be clear from now on that Popper considers that any attempt to cash out corroboration in terms of inductive probability (which obeys the laws of the classical calculus of probability) seriously misinterprets his theory of science. Though the association between those two notions seems very natural, the reader must bear in mind that, according to Popper, a good degree of corroboration never increases the statistical probability of a theory.
2. Apparently, the joint requirement of high falsifiability with good degree of corroboration cannot be met. This point is the subject of many criticisms and it defines an area where the possibility to develop satisfactorily a fully falsificationist methodology is tested. I shall return to this point later.
3. In particular, he proposed to replace the concepts of 'true' and 'false' by logical considerations about derivability relations (see *LScD*, pp. 273–4). It was a few months after the publication of *Logik der Forschung* that Popper became

acquainted with Tarski's work and learned how to respond to the current criticisms to the notion of truth. As Popper acknowledges it in several places, this radically changed his views on truth and dispelled his reluctance to use the concept.

4. Cohen makes one such attempt, though restricted to everyday reasoning. He develops a theorem that establishes how a piece of corroborative testimony might increase the support for a conclusion more than an alternative piece of testimony. See Cohen 1977, pp. 104–7.

5. Though others, such as Carnap, met this challenge by developing a theory of logical probability. I will not discuss Carnap's theory here because it is not relevant to our purposes.

6. A hypothesis is, roughly speaking, a prediction about some future state of affairs. An event, in Popper's terminology is a set of occurrences of the same kind; an occurrence is what takes place at a particular time–space region (see section 3.2.2 in Chapter 3). The attribution of probability to chancy events (e.g. the turning up of a certain face of a dice) does not seem problematic. The attribution of probability to statements that describe events (which is only a matter of terminology for people such as Reichenbach) is acceptable in some cases (e.g. the probability of the statement 'this toss will come up heads' is $\frac{1}{2}$). But, according to Popper, the attribution of probability to hypotheses (or to the statements that describe them) is not acceptable, among other reasons, because what is said with the locution 'the probability of hypothesis h is x' cannot be translated into a statement about the probability of events.

7. Reichenbach, among others, defends this claim. On his view, the assertions of natural science are in fact sequences of statements to which we must assign a probability smaller than 1. Cf. *LScD*, p. 257. It is worth noticing that while Popper has serious misgivings about restricting the interpretation of physical regularities to probability statements of relative frequency (because they satisfy the axioms of the classical theory of probability), he does not deny the meaningful use of this kind of statement in empirical science. In particular, after rejecting the subjective interpretation of probability (that treats the degree of probability as a measure of one's feelings of certainty and uncertainty, belief or doubt, on the face of certain assertions) due to its psychologistic form and being troubled by the apparent non-testable status of statements of probability in physics, he pointed out that several alternative interpretations of probability satisfy those mathematical laws as well. Accordingly, Popper developed what is known today as the propensity interpretation of probability which takes probabilities as physical propensities (considered physically as real as physical forces are) and facilitates the understanding of probabilistic physical hypotheses. This original theory, however, has little bearing on the notion of corroboration and it can be safely put aside. The reader interested in Popper's treatment of the problems of the classical theory of probability and in his own proposal should see: *LScD*, Chapter 8 and appendixes *ii–*v; *RAS*, Part II, especially Chapters 1 and 3, as well as *OU*, Chapter 4, § 27, pp. 93–5, and *WP*. For a general commentary see Ackermann 1976, Chapter 4, pp. 65–86.

8. Cf. *LScD*, pp. 257–60.

9. Ibid., p. 266; see also *OK*, pp. 18ff.

10. Falsifiability, by contrast, can be treated *always* as a purely logical affair. This should suffice to show that the parallel between falsifiability and corroboration is less than perfect, but many readers are easily misled by Popper's pervasive efforts to relate the two concepts. In fact, he says that the appraisal of the corroboration of a theory 'may be regarded as one of the logical relations between the theory and the accepted basic statements: as an appraisal that takes into consideration the severity of the tests to which the theory has been subjected' (*LScD*, p. 269). But although the severity of the tests may account for one of the aspects required to do estimations of the *degree* of falsifiability of a given hypothesis (the other being the size of the class of *PF*), it is by no means a logical notion once the test is actually carried out. It is curious that Popper does not incorporate the idea of the severity of the tests in any of his definitions of falsifiability (see sections 3.2 and 3.2.2).

11. I am leaving aside the less controversial contrast between corroborated and non-corroborated theories, since the fruitfulness of this notion reveals itself in the case of competing theories to which we attribute different positive degrees of corroboration.

12. *LScD*, p. 267 (footnote suppressed). I shall elaborate on the relation between falsifiability and corroborability later.

13. The demand for intersubjectivity guarantees the objective character of science through the public dissemination and critical discussion and correction of theories and tests. Cf. *OS*, Chapter 23 and *PH*, § 32.

14. Cf. *LScD*, p. 268. I shall refer to the attribution of negative degrees of corroboration later.

15. Notwithstanding affinities, this notion – to which Popper refers as 'absolute logical probability' in an effort to draw a sharp contrast with 'conditional probability' – should not be confused with the notion of (objective) numerical probability that is typical in the theory of games of chance and statistics, and which can be interpreted as directly proportional to logical probability. For Popper 'logical probability' is the logical interpretation of an 'absolute probability' of the sort $p(x,y)$ where y is tautological and $p(x)$ can be defined in terms of a relative (conditional) probability. Cf. ibid., pp. 118–19, 212–13, 270–73, 318.

16. Ibid., p. 269. I said earlier that the parallel between falsifiability and corroboration was imperfect, due to the different repercussions of each notion. The present discussion reveals more reasons in support of this assertion. The first difficulty that emerges is that of making precise assignments of degrees of corroboration for highly testable theories (that survive testing). All that we can say in this case is that the theory is 'well' corroborated. But of course, since subsequent corroborating instances increase the degree of corroboration very little, the assignment does not seem quite informative in the sense that once the theory has been 'well' corroborated, further attempts to corroborate it (with respect to the *same* system of accepted basic statements) appear to be futile. On the other hand, the comparison of theories according to their degree of corroboration turns out to be a very complex task and cannot be carried out without the appropriate determination of other epistemological factors; that is, their subject (whether or not they are competitors), the tests passed and their empirical content, at least.

17. Ibid., p. 275.
18. Ibid., p. 276. For the notion of levels of universality see ibid., §36, pp. 121–6.
19. For some authors the notion of criticizability (central to critical rationalism) is defective because it either leads to an infinite regress or generates some paradoxes of self-referentiality. For example, Post thinks (in 'The possible liar' and 'Paradox in critical rationalism and related theories') that for a statement to be criticizable is for it to be possibly false. But if that is the case the statement (call it R) '[e]very rational, synthetic sentence is criticizable – and this sentence is criticizable as well' is false. For R is possibly false and assuming that R is true or false we obtain: (1) That R is false entails that it is possible that R is false; (2) that it is possible that R is false entails that R is true; (3) that R is true entails that R is not false; and (4) that R is false entails that R is not false (from 1–3). From this and the contention that 'if a statement entails its negation, then its negation is necessary', Post derives the contradiction: it is not possible that R is false and it is possible that R is false (alternatively, the conclusion can be expressed in the following words: 'the statement that all rationally acceptable statements are criticizable turns out to be invalid, if it is meant to satisfy its own standard'). Bartley has met this challenge by accepting that any position which is self-referential and involves the semantic concepts of truth and falsity is liable to be found inconsistent and to produce such paradoxes (hence, the discovery of the paradoxes comes to no surprise) and by noticing (1) that critical panrationalism does not involve a theory of rationality as a property of statements; (2) that criticism does not amount to attempts to show falsity (though it may include those), and (3) that critizacibility is not justificatory. See his 'Theories of rationality' and 'A refutation of the alleged refutation of comprehensively critical rationalism', both in: Radnitzky (ed.) 1987.
20. Popper considers the distinction between logical probability and corroboration as one of the most interesting findings in the philosophy of knowledge, and the ideas of empirical content and degree of corroboration as the most important logical tools developed in his book. Cf. *LScD*, pp. 394–5.
21. Ibid., p. 418; see also p. 399.
22. It is worth noting that the logic of probability cannot solve the problem of induction. As Popper writes: 'every universal hypothesis h goes so far beyond any empirical evidence e that its probability $p(h,e)$ will always remain zero, because the universal hypothesis makes assertions about an infinite number of cases, while the number of observed cases can only be finite' (*RAS*, p. 219).
23. In a letter to *Nature* Popper and Miller have given a proof on the impossibility of inductive probability that distinguishes sharply between probabilistic support and inductive support. The proof shows that if one defines the support of h by e in terms of $S(h,e) = P(h/e)\ P(h)$ for a suitable probability function P, then one can write $S(h,e)$ as a sum $(S(h,e)) = S(h \vee \sim e, e) + S(h \vee e, e)$. It is easy to see that the first component of the sum is never positive. Since e entails $h \vee e$, Popper and Miller characterize the always non-negative support of h by e as deductive and claim that this holds for every hypothesis h and every evidence e, whether e supports h, is independent of h or undermines h. When the support of h by e is

positive (as it is if h entails e and $P(e) < 1$) then that support is purely deductive in character. The upshot of the proof is that '[a]ll probabilistic support is purely deductive: that part of a hypothesis that is not deductively entailed by the evidence [$h \vee \sim e$] is always strongly counter-supported by the evidence – the more strongly the more the evidence asserts' (Popper and Miller 1983), p. 688. In a sequel to this paper, Popper and Miller contend that though evidence may raise the probability of a hypothesis above the value it achieves on background knowledge alone, one has to attribute every such increase in probability exclusively to the deductive connections that hold between the hypothesis and the evidence (cf. Popper and Miller 1987, pp. 569–91).

24. The assignment of distinguishing marks to tested hypotheses and the subsequent comparison of them for the purposes of selection are legitimate tasks in epistemology. However, Popper holds that the first task is of limited importance because a numerical evaluation of the mark assigned to a hypothesis that has passed a test can hardly remove any difference of opinion regarding its acceptability. In fact, he writes: 'although I shall later give a definition of "degree of corroboration" – one that permits us to compare rival theories such as Newton's and Einstein's – I doubt whether a numerical evaluation of this degree will ever be of practical significance' (*RAS*, pp. 220–1).

25. Admittedly, some authors intend to refer to 'corroboration' when they speak of probability, but this does not make the two words equivalent. Popper insists on the distinction to avoid ambiguities:

> The two usages – the probability of a hypothesis with respect to its tests, and the probability of an event (or a hypothesis) with respect to its chances – have rarely been distinguished, and are mostly treated on a par. This may be due to the fact that intuitively that is to say, at least 'upon first appearances' – they are hard to distinguish. (Ibid., p. 225)

26. It may be objected that reasons beyond our control might prevent us from designing really severe tests, rendering the results of the tests actually conducted unfit to support ascriptions of corroboration. One possible reply to this objection is to leave open the possibility for others to examine critically the degree of severity of our tests and propose their own (more severe) tests. On the other hand, if the objection is meant to point out an inescapable limitation of humanly designed and conducted tests then it becomes incontestable.

27. *RAS*, p. 234 (footnote suppressed).

28. Ibid., p. 236. See also *OK*, pp. 71ff.

29. Cf. *RAS*, p. 240. Popper uses a similar resource to give a definition of the 'severity of the tests passed by a theory'. See *CR*, pp. 388–91.

30. If we use $p(e, hb) - p(e, b)$ as a measure of the degree of support given by e to h, in the presence of b (or the degree of corroboration of h by e in the presence of b) the value of $C(h,e,b)$ could be zero, which can give some support to my forthcoming suggestion about dropping any appeal to negative degrees to express corroboration.

31. The better, the more severely, a theory can be tested, the better it can be corroborated. We therefore demand that *testability and corroborability increase and decrease together* – for example that they be *proportional*. This would make corroborability inversely proportional to (absolute) logical probability ... (*RAS*, p. 245)

In the next paragraph Popper writes: 'The simplest convention will be to assume that the factor of proportionality equals 1; or in other words, that *corroborability equals testability and empirical content*.'

32. A possible exception occurs when we want to select (from a lot of refuted theories) the least bad theory, in which case we better turn to a different feature, expressed by the notion of verisimilitude, that will be discussed in the next chapter. I am leaving aside, for the sake of simplicity, the possibility of distinguishing falsified theories according to their degree of *corroboration* measured for those parts which passed severe tests.

33. This is a complex matter. On the one hand, the idea of a falsified theory (*ft*) – falsified in a particular aspect – that conserves some explanatory power and is still falsifiable in some other respect, makes perfect sense. On the other hand, if *ft* is replaced by a logically stronger competitor that explains what *ft* explained and has some extra empirical content, then we can safely ignore *ft* and devote all our attention to the successful contender – something which presumably is what happens in cases of theory-replacement. However, I will ignore the complications of this case in my further discussion.

34. We can express the same idea by saying that we want corroboration and again a high degree of *corroborability*.

35. 'A test will be said to be the more severe the greater the probability of failing it (the absolute or prior probability in the light of what I call our "background knowledge", that is to say, knowledge which, by common agreement, is not questioned while testing the theory under investigation)' (*CR*, p. 243). Note that (5) is the only way to make sense of the statement according to which theories can increase their degree of corroboration if they pass new tests. There are some complications regarding a theory that has already been accorded 1, but I believe we can safely ignore this case. On the other hand, I want to emphasize that, just like the notion of Fsb, the notion of Cr makes better sense when used as a measure to compare competing theories.

36. This shows that corroboration is not the device by which Popper can smuggle induction into his theory through the back door.

37. Though sometimes the same criticism is formulated with no reference to probability. For example, Stokes, who on this point sides with O'Hear 1980, Warnock 1960 and Watkins 1984, considers it very difficult 'to deny that corroboration involves a trace of induction, in which one accepts a statement as confirmed because of the number of severe tests that it has survived in the past' (Stokes 1998, p. 30).

38. Barker 1957, pp. 156–61.

39. But this could not be well taken because

when a scientist accepts a law, he is recommending to other men that they rely on it – rely on it, often, in practical contexts. Only by wrenching science altogether out of the context in which it really arises – the context of men trying to change and control the world – can Popper even put forward his peculiar view on induction. Ideas are not *just* ideas; they are guides to action. Our notions of 'knowledge', 'probability', 'certainty', etc., are all linked to and frequently used in contexts in which action is at issue: may I confidently rely upon a certain idea? Shall I rely upon it tentatively, with certain caution? Is it necessary to check on it? (Putnam 1974, p. 222)

40. Ibid., p. 225. Alfred Ayer makes a very similar point: 'whereas [the inductivists] hold that the accumulation of positive instances confirms a universal hypothesis, he [Popper] prefers to say that the failure of the instances to be negative corroborates it. I confess that I have never been able to see that this difference amounted to very much' (Ayer 1974, p. 685).
41. Putnam defines the term 'theory' in terms of a 'set of *laws*' and understands these as 'statements that we hope to be *true*; they are supposed to be true by the nature of things, and not just by accident', drawing a sharp distinction with regard to boundary conditions, which he takes as usually false (cf. Putnam 1974, p. 226).
42. Ibid., p. 225. Putnam contends that Popper ignores the role of AS.
43. Ibid., p. 227 (emphasis added).
44. Putnam spends about one third of his paper explaining Kuhn's views and examining the dispute between Popper and Kuhn. Since Putnam's strictures of corroboration do not depend on this topic, I will set it aside.
45. But this is not the sense in which Popper uses the notion of logical (im)probability. With this notion Popper usually indicates content as a measure of the class of potential falsifiers, and Putnam finds it troubling that this measure cannot be expressed as a cardinality (see Chapter 3, above).
46. Putnam 1974, p. 237.
47. *LScD*, note *1 to section 28, p. 101.
48. Cf. *RAS*, p. 256. For another counter-instance to Putnam's accusation, see also *CR*, p. 385.
49. Popper 1974, p. 995.
50. 'The old scientific ideal of *epistēmē* – of absolutely certain, demonstrable knowledge – has proved to be an idol. The demand for scientific objectivity makes it inevitable that every scientific statement must remain *tentative for ever*. It may indeed be corroborated, but every corroboration is relative to other statements which, again, are tentative. Only in our subjective experiences of conviction, in our subjective faith, can we be "absolutely certain"' (*LScD*, p. 280). For additional reasons to hold scientific theories provisionally see Simkin 1993, p. 19.
51. At this point Lakatos makes a disclaimer: 'of course it is abundantly clear that Popper's *instinctive* answer [in the days of *Logik der Forschung*] was that the aim of science *was* indeed the pursuit of Truth' (Lakatos 1974, p. 253).
52. This is mainly because he lacks an adequate theory of truth. So,

> *Popper's demarcation criterion has nothing to do with epistemology.* It says nothing about the epistemological value of the scientific game. One may, of course, *independently* of one's logic of discovery, *believe* that the external world exists, that there are natural laws and even that the scientific game produces propositions ever nearer to Truth; but there is nothing *rational* about these metaphysical beliefs; they are mere animal beliefs. There is nothing in the *Logik der Forschung* with which the most radical sceptic need disagree. (Ibid., p. 254, footnote suppressed.)

53. Ibid. For an exposition of the Tarskian turn and an explanation of verisimilitude see Chapter 5 below.
54. Ibid., p. 256.
55. Ibid. Popper's phrase comes from *CR*.
56. As any reader of *LScD* can tell, section 84 concerns the use of the concepts 'True' and 'Corroborated'. Popper writes that although it is possible to avoid using the concepts 'true' and 'false' we are not forbidden to use them. See also the first Addendum to *OS*.
57. See Popper 1974.
58. I hope the reader will indulge me for not saying more about verisimilitude since the next chapter deals entirely with this notion. In the meantime, let me mention that Popper distinguishes between two questions: the meaning of the statement 'The theory t_2 has a higher degree of verisimilitude than the theory t_1' and 'How do you know that the theory t_2 has a higher degree of verisimilitude than the theory t_1'? The latter can be answered only after answering the former

   > and it is exactly analogous to the following (absolute rather than comparative) question about truth: 'I do *not* know – I only guess. But I can examine my guess critically, and if it withstands severe criticism, then this fact may be taken as a good critical reason in favour of it.' (*CR*, p. 234)

59. Derksen 1985, pp. 329–30. Derksen restates the point in this way: 'What reason have we to believe that the distribution of the theory's implications about its truth and falsity content will stay so favourable for us? What reason have we to believe (or, if you want, to guess) that the theory will apply then as well as it did so far? The reason that comes to mind, viz. that the future is like the past – to use the traditional formulation – is so inductivistic that it would even make a complacent inductivist shiver.' Since Popper is not a justificationist, he could simply reply that we do not have any reason to guess (not to believe) that a report on corroboration may indicate certain distributions of truth-content and falsity-content. But the Popperian need not worry about those distributions staying favourable to us, because the theory can easily accommodate (since it anticipates them) positive as well as negative cases.
60. Ibid., p. 325.

5
Verisimilitude

> I should perhaps stress that, to my mind, Popper is the only person who has made the slightest progress towards solving the problem of verisimilitude, a problem that he himself more or less discovered.
>
> Miller (1974)

As explained in the previous chapter, Popper's theory of science can be formulated without appealing to the notion of truth and without falling into pragmatism or instrumentalism. This special feature, however, is independent of the identification of the aim of empirical science in terms of a general search for the truth, a characterization that Popper considers as a trite idea. In *LScD*, he makes several remarks to that effect, when explaining the distinction between the concepts 'true' and 'corroborated', but it will not be until *CR* that he presents his theory of verisimilitude. Roughly speaking, degree of verisimilitude expresses the relative distance to the truth of a (false) statement or a (partially false) theory. This notion is important because it makes it possible to define progress in a series of false theories.[1] Assuming that theories are verisimilar (instead of true *simpliciter*), it follows that they neither completely miss the truth, nor reach the truth. The aim of science can be promoted by formulating theories that are closer to the truth than their overthrown competitors. But the best way to estimate the relative distance to the truth of a particular theory is through a comparison with a genuine rival. Given two competing theories, it is the task of the scientist to decide which of them constitutes a better approximation to the truth. In this chapter, I shall introduce Popper's notion of verisimilitude, discuss its role in a falsificationist epistemology, consider the main criticisms of it, and propose a response to objections against Popper's qualitative definition of 'verisimilitude'.

5.1 The theory of objective truth

No theory of empirical science would be satisfactory without a notion of its progress. So far, we have explored ways of demarcating science from pseudo-science, of selecting between competing hypotheses and of securing the empirical character and logical strength of scientific theories. But, although we have given some criteria to select the best theory from several, we have not said anything yet about what constitutes progress in the endless game of science.[2] Popper's notion of verisimilitude would complete the picture, because now we can say that science grows by making successively better approximations to the truth. In other words, progress in science takes place when theories with a lesser degree of verisimilitude are overthrown by theories with a greater degree of verisimilitude. Why, however, are we interested in verisimilitude rather than being concerned with truth? And how can we make sense of the idea that science progresses by successive approximations to the truth? Before attempting to answer these questions, we need to understand the role of truth in Popper's epistemology. Popper subscribes to Tarski's correspondence theory of truth. On this view, a statement is true if the proposition it expresses accords with the corresponding state of affairs. For instance, the statement 'snow is white' is true if and only if snow is white; the statement 'grass is red' is true if and only if grass is red. Strictly speaking, only the truth of certain empirical statements can be conclusively obtained. By contrast, theories (i.e. sets of universal statements), in general, cannot be true,[3] but can have different degrees of verisimilitude, which means that they are closer or farther from the truth. According to Popper, the great merit of Tarski's work is that it rehabilitated the old intuitive idea that absolute, objective truth is correspondence with the facts, and enabled us to use this notion in epistemology without running into logical difficulties.[4]

In *LScD* Popper formulated his theory of science with no appeal whatsoever to the notion of truth. He avoided using this notion in his epistemology and thought it possible to replace it by deducibility and similar logical relations. In *CR* (and later on in Chapter 9 of *OK*) he explained his reluctance to use the notion of truth as a result of his finding it problematic, and especially as a result of not being able to give a satisfactory formulation of the correspondence theory nor having an answer to the criticisms of it, embodied in the semantic paradoxes. All this changed radically after Tarski's work. Popper embraced his theory enthusiastically as soon as he became acquainted with it because, according to him, Tarski legitimized the use of an absolute or objective notion of truth defined as correspondence. So far as I know, all that matters for Popper's theory of verisimilitude is the availability of an objective theory of truth that does justice to the old intuition of correspondence between statements and states of affairs.

According to Popper, Tarski's theory of truth fulfils that desideratum. Properly understood, it provides us with good reasons to avoid slipping into conventionalism and from that into relativism. But it does not provide a general criterion of truth in the sense that it is not a theory of how to discover the truth.[5] Another important consequence of Tarski's theory – Popper holds – is that it affords a way to overcome the mistaken dogma that a satisfactory theory of truth would have to be a theory of true belief (or well-founded or rational belief). This result comes in very handy for Popper's theory of science, since he rejects a subjectivist epistemology in all its variants. Indeed, he thinks that they confuse consistency with truth, 'known to be true' with truth, pragmatic utility with truth, and that they reduce knowledge to beliefs and mental states. We can see that these identifications are wrong by noting that some statements may be true regardless of whether anybody believes them and, conversely, that some others may be false despite universal belief to the contrary. Presumably, such problems are absent in an objective epistemology that deals exclusively with the logical content of our theories, conjectures or guesses. Accordingly, we can assert that 'a theory may be true even though nobody believes it, and even though we have no reason for accepting it, or for believing that it is true; and another theory may be false, although we have comparatively good reasons for accepting it'.[6]

In Popper's epistemology, truth becomes a regulative principle and it plays the role of guiding our search for better theories under the assumption that we might learn from our mistakes. Popper expresses this idea with the metaphor of a mountain peak that is always – or almost always – wrapped in clouds, but which nonetheless is the inspiring goal of a certain kind of climber. The fact that the peak cannot be seen does not preclude its regulative force any more than repeated failures to reach it discourage the truly motivated man from making additional attempts, avoiding the unsuccessful paths he formerly took but using the ones that gave him some success:

> The climber may not merely have difficulties in getting there – he may not know when he gets there, because he may be unable to distinguish, in the clouds, between the main summit and some subsidiary peak. Yet this does not affect the objective existence of the summit, and if the climber tells us 'I have some doubts whether I reached the actual summit', then he does, by implication, recognize the objective existence of the summit. The very idea of error, or of doubt (in its normal straightforward sense) implies the idea of an objective truth which we may fail to reach.[7]

In a similar way, we can make progress towards the truth even though we cannot say exactly where the truth is. By contrast, there are cases in which we can be sure that we have not reached the truth; hence, we are in a position to dismiss the theory that we were considering as a candidate for reaching that regulative goal. The former cases can be dealt with by allowing that there are degrees of verisimilitude (relevant to better or worst correspondence) that can be sensibly predicated of a pair of theories, say t_1 and t_2, so that one supersedes the other because it approaches more closely the truth. Popper finds this way of referring to the standing of theories with regard to the truth (in the Tarskian sense) illuminating for the purposes of understanding how a theory can correspond to the facts better than a competitor.

5.2 Popper's notion of verisimilitude

Popper presented his original definition of 'verisimilitude' in a lecture entitled 'Truth, rationality, and the growth of scientific knowledge' partially delivered in the International Congress for the Philosophy of Science (Stanford, 1960) and included now as Chapter 10 of *CR*. The concept is cashed out in terms of the notions of truth and content, and Popper claims we might obtain compatible results using either logical or empirical content. By using the Tarskian idea of logical consequence, he defines the content of a statement a, as the class of all its logical consequences. Of course, if a is true, all its logical consequences will be true (since truth is always transmitted from a premise to all of its conclusions); if a is false, then we may encounter both true and false statements in a's consequence class. The subclass of all true statements in a's consequence class is called the 'truth content' of a; the subclass of all false statements in a's consequence class is called the 'falsity content' of a.[8] Assuming that both the truth content and the falsity content of a theory a can be measured in principle, Popper defines the verisimilitude (Vs) or truthlikeness of a in the following way:

(1) $$Vs(a) = \boldsymbol{Ct}_T(a) - \boldsymbol{Ct}_F(a)$$ [9]

Where $\boldsymbol{Ct}_T(a)$ is a *measure* of the truth content of a, and $\boldsymbol{Ct}_F(a)$ is a *measure* of the falsity content of a. *Prima facie*, this definition seems to satisfy our intuitions. For it makes a theory's degree of closeness to the truth a function of both the measure of its truth content and the measure of its falsity content. The degree of verisimilitude of a theory t increases with its truth content and decreases with its falsity content in such a way that if we could add to its truth content whilst subtracting from its falsity content we would be improving the theory.[10] Ideally, a theory with an empty subclass of false statements that speaks about all the facts of the world would have the maximum degree of verisimilitude and should be deemed absolutely comprehensively true,[11] yet such a perfect theory does not seem to be obtainable (as we will see later) if the principles of Popper's philosophy of science are right. On the other hand, since empirical science progresses by conjecture and refutation, that is, by the continuous replacement of false theories by competitors that are better fitted to survive testing, the definition expressed by (1) proves its fertility when we use it to make comparisons of those theories' respective degrees of verisimilitude. For if we could find a theory endowed with higher truth content and lower or equal falsity content than one of its rivals, then such a theory would be closer to the truth, as we can see in the following application of (1).

Consider theories t_1 and t_2, and assume that their respective truth contents and falsity contents are comparable. Then we can say that t_2 is a better approximation to the truth or that it is more closely similar to the truth than t_1, if and only if (a) t_2 has greater truth content than t_1, but does not have greater falsity content than t_1, or (b) t_1 has greater falsity content than t_2, but does not have greater truth content than t_2.

(2) $Vs(t_2) > Vs(t_1)$ iff $(\boldsymbol{Ct}_T(t_2) > \boldsymbol{Ct}_T(t_1)$ and $\boldsymbol{Ct}_F(t_2) \leq \boldsymbol{Ct}_F(t_1))$,

or: $(\boldsymbol{Ct}_T(t_1) \leq \boldsymbol{Ct}_T(t_2)$ and $\boldsymbol{Ct}_F(t_2) < \boldsymbol{Ct}_F(t_1))$.

There are, however, several difficulties with the definitions in (1) and (2) above. Since (2) amounts to the application of (1) to a comparison between any pair of self-consistent, different and genuinely competing theories, it is possible to illustrate the difficulties by restricting our analysis to (1).[12] This should suffice to make the point and is easier to accomplish. If the definition is correct, then a theory t_2 comprised of a single true statement and having an empty subclass of false statements would be better than a competitor (call it t_1) that has two statements – one true and the other false – hence a non-empty subclass of false statements. Now, the natural candidates for t_2 are tautologies and trivial statements such as Popper's example, 'snow is usually white'. But on a closer look, the latter seems to have very little informative content, and the former type of statement is completely devoid of informative content, according to Popper's principles. Moreover, all statements in the consequence class of a tautology must be tautologies; hence no tautology can have a non-empty falsity-content subclass. Thus, according to (1) any tautology should exceed t_1 by verisimilitude, but Popper's theory of empirical science rightly excludes tautological truths as the goal towards which science aims. On the other hand, many trivial statements (e.g. isolated existential statements) are not falsifiable. Hence they do not have empirical content (or such content is negligible). Therefore, no theory whose only true statement is one of the kind just mentioned can be considered for a comparison in terms of verisimilitude since, strictly speaking, it is metaphysical.

In the addenda to *CR*, Popper offered a few additional distinctions that give a better definition of 'verisimilitude' and ameliorate the difficulties posed by tautological statements. Without bringing in all the technical details, I shall mention the most important distinctions introduced and briefly explain what they contribute to the notion of verisimilitude in the framework of a falsificationist theory of *empirical* science. To begin with, we need to define truth content in such a way that (a) the concept can be applied also to false theories and (b) it yields a zero value when applied to formal truths or necessarily true statements (since we want to rule out the view that empirical science aims for tautological truth). Let T denote the class of all true statements of some language L, and $\boldsymbol{Ct}$ stand for the measure of (logical) content, then we can write:

(3) $a \in T \rightarrow \boldsymbol{Ct}_T(a) = \boldsymbol{Ct}(a)$

which means that if statement *a* is true, then (the *measure* of) its truth content is equal to (the *measure* of) its logical content. It is possible to express the logical content in terms of the calculus of probability by writing: $Ct(a) = 1 - p(a)$.[13] On the other hand, since every false statement entails a class of true statements, we are justified in attributing truth content to it. Suppose b is a false statement. In this case $Ct_T(b)$ is a subclass of $Ct(b)$ (hence, $Ct_T(b) < Ct(b)$), since the latter contains both false and true statements. If a is a tautology (*tautol*), then the probabilistic formula above will give us a value of 0 (assuming we attribute maximal logical probability to tautologies and express that by 1). For a self-contradictory system (*contrad*), we would obtain a probability value of 1 (under parallel assumptions), but Popper wants to reserve this value for maximal truth content, so we need to make some changes. To have a satisfactory characterization of falsity content Popper lays down a number of desiderata. Let F stand for the class of all false statements of some language L. Thus

(4) (i) $a \in T \rightarrow Ct_F(a) = 0$

(ii) $a \in F \rightarrow Ct_F(a) \leq Ct(a)$

(iii) $0 \leq Ct_F(a) \leq Ct(a) \leq 1$

(iv) $Ct_F(contrad) = Ct(contrad) = 1$

(v) $Ct_T(a) = 0 \rightarrow Ct_F(a) = Ct(a)$

(vi) $Ct_F(a) = 0 \rightarrow Ct_T(a) = Ct(a)$

(vii) $Ct_T(a) + Ct_F(a) \geq Ct(a)$[14]

Some of these desiderata are fairly obvious under the assumption that the maximal value of truth content in an infinite universe should be greater than 0 and equal to 1. According to Popper, desiderata (iv) and (vii) are to make sense of the following intuitive consequences: (a) since everything (including all truths) follows from a contradiction, then the measure of the content of a contradiction equals 1; (b) we expect that the measure of the falsity content of a contradiction be at least equal to the measure of its truth content. Without (iv) and (vii) we would arrive at the unwelcome result that in most cases the falsity content of a contradiction will be smaller than its truth content. To remove these blemishes, Popper replaces desideratum (iv) by (iv, a) $Ct_F(contrad) = $ constant, and (iv, b) $Ct_F(contrad) \leq Ct_T(contrad)$. Since verisimilitude (see (1) and (2) above) is expressed as a function of both truth content and falsity content, we finally arrive at the following results:

(5) $$Vs(tautol) = 0$$
$$Vs(contrad) = -1$$
$$-1 \leq Vs(a) \leq +1$$[15]

Before moving on to the next section, let me stress that Popper has worked out the definitions in such a way that we accord a zero degree of verisimilitude to tautologies. To obtain this result, he has appealed to the calculus of probability combined with his idea that the logical probability of a statement is inversely proportional to its empirical content. But it must be clear that attributing a greater (or lesser) degree of verisimilitude to a theory is a completely different affair from attributing a higher (or lesser) degree of probability to it; and that verisimilitude should not be confused with probability. Some philosophers have confused verisimilitude and probability because both notions are related to the concept of truth and they both introduce the idea of approaching the truth by degrees. However, one can see the difference clearly, at least for the case of *logical* probability, since this represents the idea of approaching logical certainty (tautological truth) through a gradual diminution of informative content (thus, it combines truth with *lack* of content), whereas verisimilitude entails the possibility of approaching comprehensive truth by combining truth with content.

5.3 Refining verisimilitude

The sort of difficulties mentioned above might have led Popper to restrict the type of statements that could enter the truth-content subclass of a theory. In the new definition formulated in *OK* he explicitly rules tautologies out of this subclass. Popper explained the motivation for introducing the new qualification to the definition of 'verisimilitude' in the following way:

> Every statement has a content or consequence class, the class of all those statements which follow from it. (We may describe the consequence class of tautological statements, following Tarski, as the zero class, so that tautological statements have zero content.) And every content contains a sub-content consisting of the class of all and only all its *true* consequences. The class of all the *true* statements which follow from a given statement (or which belong to a given deductive system) and which are not tautological can be called its *truth content*.[16]

By courtesy, the class of false statements entailed by a statement can be called its falsity-content subclass.[17] Notice that, according to Popper's definition, the idea of verisimilitude applies equally to statements or theories. Then, the foregoing caveat will preclude comparisons where one deals with at least one statement (or theory) that is tautological, since its logical content will be comprised only of tautologies and, *a fortiori*, its 'truth-content' subclass (quantified as zero) would be empty. This exclusion, of course, is consistent with

Popper's main definitions, according to which the empirical (informative) content of metaphysical theories as well as of tautological ones is equal to zero, since their respective logical content is devoid of empirical import. From this it follows that they have neither truth content nor falsity content. On the other hand, all empirical statements and theories, including false ones, possess a truth content greater than zero, because, even being false, they are at some distance (no matter how big) from the truth, or as Popper suggests, they have some resemblance to the truth (some degree of verisimilitude). In the light of this, we need to reinterpret (2) in such a way that truth content is to be restricted to nontautological consequences, as follows:

(2*) $Vs(t_2) > Vs(t_1)$ iff $(\boldsymbol{Ct}_T(t_2) > \boldsymbol{Ct}_T(t_1)$ and $\boldsymbol{Ct}_F(t_2) \leq \boldsymbol{Ct}_F(t_1))$,

or $(\boldsymbol{Ct}_T(t_1) \leq \boldsymbol{Ct}_T(t_2)$ and $\boldsymbol{Ct}_F(t_2) < \boldsymbol{Ct}_F(t_1))$

(Where 'Ct_T' is *free* of tautologies)

Let us use a topological representation of the relative standing of different theories according to their respective degree of verisimilitude. If we follow Popper's definitions, this should be the right representation (I let 'ACTW' stand for the absolute comprehensively true theory of the world):

$$
\begin{array}{c}
Vs \\
(6) \quad -1 \quad\rule[0.5ex]{10em}{0.4pt}\quad 0 \quad\rule[0.5ex]{10em}{0.4pt}\quad +1 \\
\text{Self-contradictions} \quad \text{Tautological theories} \quad \text{Falsified theories} \quad \text{ACTW}
\end{array}
$$

However, I have some misgivings with regard to what (6) offers us. To begin with, I do not see any reason to include in the diagram self-contradictions and tautologies (though one might want to include them to indicate the outside boundaries as done in the diagram for *Fsb*). If we are concerned with the idea of approaching absolute empirical truth, it is perfectly appropriate to focus all discussion on empirical theories. Moreover, Popper insists that the main point of verisimilitude is the possibility of making comparisons, and that it is only through comparison that the idea of a higher or lower degree of verisimilitude becomes applicable for the analysis of the methods of science. Of course, no comparison with respect to self-contradictions or tautologies seems to be in place here. Consequently, I think we can exclude from the diagram these theories with no risk of affecting our investigation of verisimilitude. Secondly, the only reason to accord some degree of verisimilitude to a contradictory theory is that everything follows from it. In my opinion, this is just a peculiarity of deductive systems (i.e. a peculiarity of logic), but we need not transfer it to our epistemology. The minimum requirement we are entitled to demand of a scientific theory is that it is not self-contradictory; hence, this seems to provide additional support to my suggestion of excluding this type of theory from (6). A simpler, more adequate representation of the degree of verisimilitude of various falsified (false) theories would be:

$$(7) \quad 0 \underset{\text{ACTW}}{\overline{\qquad ft_1 \qquad\qquad ft_2 \qquad\qquad ft_3 \qquad\qquad ft_4 \qquad\qquad ft_5 \qquad}} 1$$

I make no attempt to suggest a diagram for empirical statements. Firstly, because if they are true, it does not seem appropriate to suggest that they might have one or other degree of verisimilitude. When considering ascriptions of verisimilitude to *true* statements, one may have in mind that they provide information about the world with differential degrees of precision, universality and explanatory power, and that these features ground one's decision to rank them accordingly. But such features pertain more to theories than to statements, therefore that claim seems to be misconceived. On the other hand, comparing true statements with different degrees of precision or universality is usually done with regard to those very features (precision and universality) and it is hard to make sense of any attempt to establish a comparison between them in terms of verisimilitude, without conflating the two notions.[18] False statements are quite different, since they have truth content according to the definitions above. Popper has managed to illustrate how to attribute different degrees of verisimilitude to a pair of false statements by expressing them as interval statements in such a way that they admit a consecutive range of values, that is, a range of error.[19] This procedure brings in surreptitiously the two conditions that seem to be necessary for meaningful talk about verisimilitude: (i) that the compared items are genuine rivals; and (ii) that we have some way to rank order them according to their closeness to the truth. Again, we see how crucial is the feature of comparability for a proper understanding of verisimilitude.

A comparison is something that can be performed properly amongst distinct objects that nonetheless share some feature (otherwise no meaningful comparison could be established). Excluding the case of absolute comprehensively true theories, where talk about verisimilitude is out of the question (since they are not truthlike, but *true*), and any comparison regarding this property should yield the result that they are equally good (there is no sense in choosing between two rival comprehensively *true* theories, and one may even expect that this is just not a possible epistemological situation) the comparison between theories produces interesting results when we have to deal with false theories (again, which are not exactly at the *same* distance to the truth). Can we meaningfully claim that there are some falsehoods less false than other falsehoods? This must be the right conclusion to draw, since the Popperian story about the growth of science privileges a process in which one false theory is discarded in favour of another (potentially) false theory. If this is a step towards progress, it must make sense to say that the new theory, although also false, is closer to the truth than its discarded predecessor. But even if we are comparing a pair of falsified theories, it is still possible to determine which one is better by considering the tests that each has passed. Suppose, for example, that t_2 has withstood tests which t_1 failed, then this 'may be a good indication that the falsity content of t_1 exceeds that of t_2 while its truth content does not. Thus we may still give preference to t_2, even after its falsification, because we have reason to think that it agrees better with the facts than did t_1.'[20]

It is important to mention at this point that ascriptions of verisimilitude have the same conjectural status that we attributed before to estimates of degree of corroboration and even to attributions of degree of falsifiability. They all involve (tentative) appraisals, and though we should expect that the relative appraisal of two competing theories, t_1 and t_2, will remain stable in the absence of radical changes in our background knowledge, it is not impossible that the detection of mistakes, the dictates of rational criticism or the discovery of new facts overturn it, and consequently that we have to revamp our appraisal. Despite this possibility, however, we can always appeal to a theory's content and explanatory power to justify our preference for it and ground our conjecture about its truthlikeness.

Before I move on to the most important objections against verisimilitude, let me mention that in order to make sense of the idea of progress towards the truth we need to check the compliance of at least three requirements in the theory to which we attribute better agreement with the facts. According to Popper, these requirements (in light of the previous theories) are: (i) Simplicity. The better theory should aim to explain the structural properties of the world; it should proceed from some simple, new and unifying idea and be capable of connecting things or facts that were not related before. (ii) Testability. The new theory must be independently testable. Besides explaining all the facts that its older competitor did, the new theory must have excess empirical content, that is, it must make new (negative) predictions. (iii) Empirical success. The new theory must pass some new and severe tests; it must have a better degree of corroboration than its competitors. Now, the nature of the last requirement is clearly different from the nature of (i) and (ii). While these are logical (and so merely formal) requirements that are needed to decide whether the newly advanced theory should be taken seriously and considered as a good candidate for empirical testing, (iii) is a material requirement that can only be checked in the empirical realm and need not be met, for even if the theory fails new tests (from its outset) it can be still considered as a good approximation to the truth.[21] For this reason, (iii) can be waived except when we are interested in increasingly approaching the truth, in which case

Verisimilitude   131

we demand also positive successes, for science would stagnate if we did not obtain the corroboration of new, interesting predictions. All this can be summed up by saying that if we accept truth as the regulative idea of science, we should make sure that our new theories (that make successful predictions) have less falsity content and more truth content than their competitors: that is, that they have a greater degree of verisimilitude.

5.4 Criticisms of Popper's notion of verisimilitude

Popper's notion of verisimilitude has attracted criticism from several quarters. The most general objection (and I would say, the most indirect, since Popper does not endorse the view) points out the inconvenience of considering as the aim of science the search for verisimilitude instead of the search for truth. Another common objection branches from the potential links between verisimilitude and corroboration. In the previous chapter, we saw that Lakatos confuses corroboration with a *measure* of verisimilitude and that he takes the (alleged) difficulties of the former notion equally to affect the latter. But it should not be difficult for the reader to clear this point up and dismiss Lakatos's qualms, since we have already introduced an articulated formulation of the measure of verisimilitude that is completely independent of corroboration.[22] A favoured criticism charges that the stability of ascriptions of verisimilitude presupposes induction (Lakatos, Derksen), but this quandary does not seem to be really on the track if we consider that such appraisals have also a conjectural character and are not advanced as predictions about the future values that we may accord to a particular theory. Furthermore, this criticism can be met by stressing that there is a perfect non-inductive connection between falsification and truth: namely, that a hypothesis that contradicts a true test-statement is false; whereas one that (so far as we know) does not, may be (for all we know) true. The most serious criticisms, however, have been directed at the definition of 'verisimilitude', and charge that neither the quantitative nor the qualitative definition provide good grounds to attempt ascriptions of verisimilitude that serve the purposes this notion is supposed to have in Popper's theory of science. Critics contend that the quantitative definition yields all kinds of unwelcome results, such as assigning a higher degree of verisimilitude to theories which intuitively should receive a lower mark (for example, because they have less empirical content), thus failing to discriminate adequately between two unequal competitors. But the most devastating criticism is levelled against the qualitative definition. Roughly speaking, it argues that this definition is unable to distinguish between two false theories (which in fact are not equally close to the truth) because it cannot assign to them different degrees of verisimilitude. The champions of this objection are

Tichý and Miller who have argued, independently, that the definition in (2) does not work if our aim is to compare unequally false theories expecting to find discriminating measures of verisimilitude because (2) forces us to assign an *equal* degree of verisimilitude to any pair of false competing theories which are not at the *same* distance from the truth. In what follows, I shall explain Tichý's and Miller's criticisms and propose an amendment to the qualitative definition of 'verisimilitude' that will allow us to circumvent the charge just mentioned and block their objection.

5.4.1 *Tichý: all false theories are equally verisimilar*

Tichý contends that for very simple logical reasons, Popper's definitions of verisimilitude are totally inadequate. Let us recast his criticism of (2) by giving first the basic logical definitions of it. He calls any finite set of (closed) sentences of a language L (that has a finite number of primitive descriptive constants) a theory of L. Suppose $A, B, C \ldots$ are arbitrary theories in L. Let $Cn(A)$ stand for the set of logical consequences (theorems) of A, and T and F for the set of true and false sentences of L, respectively. Then he can rewrite Popper's definitions (1) and (2) accordingly. He writes A_T and A_F for the truth content and the falsity content of A, and assumes that the truth contents and the falsity contents of two theories, A and B, are comparable just in case one of them is a proper (or improper) subclass of the other. Then, according to Popper's logical definition of V_s,

    A has less verisimilitude than B just in case

    (a) A_T and A_F are respectively comparable with B_T and B_F, and

    (b)   either $A_T \subset B_T$ and $A_F \not\subset B_F$ or $B_T \not\subset A_T$ and $B_F \subset A_F$.[23]

Conjoining this with the requirement for comparability above, we obtain the result that A has less verisimilitude than B just in case either $A_T \subset B_T$ and $B_F \subseteq A_F$ or $A_T \subseteq B_T$ and $B_F \subset A_F$. According to Tichý, the foregoing (as well as (2)) fails to give an adequate explication of verisimilitude because if B is false we can never obtain the result that A has less verisimilitude than B, in clear opposition to what was intended. Tichý supports this claim with the following argument. If B is false, then there is at least one false sentence (call it f) in $Cn(B)$. Assume that $A_T \subset B_T$. In this case, there is at least one sentence (call it b) which is in B_T but not in A_T. However, $(f \cdot b) \in B_F$, whereas $(f \cdot b) \notin A_F$, since we have identified b as accounting for the excess truth content of B over A (which means that $b \notin A_T$). Therefore, $B_F \not\subset A_F$, which violates one of the conditions just stated. We obtain a similar result assuming that $B_F \subset A_F$. In this case, there must be at least one sentence (call it a), which is in A_F but not in B_F. However, $(f \supset a) \in A_T$, whereas $(f \supset a) \notin B_T$, since a does not belong to the falsity content of B. Therefore, $A_T \not\subset B_T$.

In the second part of his paper, Tichý argues that given two theories A and B, such that A is patently closer to the truth than B, Popper's probabilistic definition would attribute a lesser degree of verisimilitude to A.[24] In this part, Tichý starts by considering an example. Suppose we have a rudimentary weather-language L with no predicates and only three primitive sentences, 'it is raining', 'it is windy' and 'it is warm' (further abbreviated as 'p', 'q', and 'r', respectively). If we assume that all three sentences correspond to the facts, then writing t for $p \cdot q \cdot r$ (which is a true theory of L) we obtain $T = Cn(t)$. Next, we are asked to consider the *constituents* of this theory (the eight sentences that start with t and gradually replace each component by its negation, exhausting all logical alternatives). These sentences are mutually incompatible, jointly exhaustive and of equal logical strength (the logical probability of each is 1/8). Given the relations of compatibility and incompatibility for any pair of sentences a and b of L, the definition of 'true in L', and the fact that every consistent sentence a of L is logically equivalent to a disjunction of constituents (the disjunctive *normal form* of a) Tichý proves that if a is false then $ct_T(a) = (7/8) - p(a)$ and $Ct_F(a) = 1 - [p(a)/p(a) + 1/8]$.

The problem – he continues – is that, according to the definitions, the values of verisimilitude of false sentences of L depend only on their logical probabilities in such a way that no factual knowledge over and above the knowledge that the two theories are false would be required to decide which of the two theories is closer to the truth. On the other hand, 'we want it to be possible for one false theory to be closer to the truth than another false theory despite the two theories having the same logical probability'.[25] Now suppose that Jones and Smith (who share a windowless and air-conditioned cell in prison) use L to discuss the weather. Jones submits the theory: $\sim p \cdot \sim q \cdot \sim r$ (he thinks 'it is a dry, still day, with a low temperature'), whereas Smith thinks that the temperature is low but conjectures (rightly) that it is raining and windy. Smith's theory is $p \cdot q \cdot \sim r$. It seems undeniable in the example that Smith's theory is a better approximation to the truth than Jones's. Whereas Jones is wrong in all three counts (and therefore could not be farther from the truth), Smith is wrong only as far as temperature is concerned (just on one count). One should expect, then, that Smith's theory has a higher degree of verisimilitude: that it exceeds Jones's theory regarding the measure of truth content and, consequently, that Jones's theory exceeds Smith's theory regarding falsity content. But the values[26] assigned to each of these functions are the same for both theories.

The situation gets worse when we consider a theory that is potentially farther from the truth than another theory, because in some of these cases the probabilistic definitions of verisimilitude accord a greater degree of verisimilitude to that theory and do a poor job estimating its measure of falsity content, by contrast with the corresponding measure for its more truthlike competitor.

Suppose now that Smith sticks to his theory (call it S) while Jones weakens his claim by dropping his (wrong) conjecture about the temperature. He maintains now that it is a dry and still day ($\sim p \cdot \sim q$ – call this theory J) and withholds judgement on the third fact. Since he is now wrong only on two counts, his new theory is marginally better than his old one, but it is still inadequate and not good enough to match (let alone supersede) Smith's theory. One should expect both that Jones's theory will exceed Smith's in falsity content and also have a lesser degree of verisimilitude.

However, according to the tabulated values, $Ct_F(J)$ is strictly less than the $Ct_F(S)$ and $Vs(S)$ is strictly less than the $Vs(J)$. Tichý concludes that Popper's definitions of verisimilitude are wanting, but he suggests a solution for a simple language based on propositional logic, that supposedly meets all the intuitive requirements. First, we define the 'distance' between two constituents as the number of primitive sentences negated in one of the constituents but not in the other. Then we define the verisimilitude of an arbitrary sentence a as the arithmetical mean of the distances between the true constituent t and the constituents that appear in the disjunctive normal form of a.[27]

I shall deal with Tichý's criticisms in due course. Before proceeding to Miller, let me point out that his suggestion that we want to be able to distinguish by verisimilitude two false theories with *equal* logical probability seems to be plainly wrong (I am assuming he means by that logical strength, that is logical *improbability*). If, as Popper's theory maintains, any epistemologically interesting increase in verisimilitude amounts to a corresponding increase in logical content, then what Tichý envisages can never be the case nor does it correspond to what the Popperian wants. Quite the contrary: one should expect that any two false theories that have the same logical strength would have the same degree of verisimilitude, hence they would not be comparable (or the comparison would not be meaningful).

5.4.2 *Miller: no two false theories are comparable by verisimilitude*

According to Miller, the problem of verisimilitude investigates what can there be about one false theory that makes it closer to the truth than another false theory. In his opinion, the answer to this question has to do with the notions of logical strength, power or informative content of a deductive theory. Intuitively speaking, one false theory, say t_2, can be better than another false theory, say t_1, if t_2 is logically stronger than t_1, and t_2 says more than t_1.[28] Miller restricts his criticisms to the qualitative definition of 'verisimilitude' – which he considers to be 'quite untenable' – and explains that when we are dealing with a 'purely qualitative' theory of verisimilitude, for comparisons of truth content (or falsity content) we require that the measures be so related that one includes another (in the usual set-theoretical sense). That many pairs of theories turn out not to be comparable either by truth content or by falsity content (on Popper's theory) is a consequence of that feature of the qualitative definition.

Miller contends that no two false theories are comparable for verisimilitude under Popper's definitions of this notion; moreover, he insists that both definitions are operative only for true theories (though they might work when there is at least one true theory involved). To better understand Miller's point let us quickly look to his more important logical stipulations. To begin with, he treats theories as formulated in some first-order language or other, and assumes that no consistent theory is finitely axiomatizable. As expected, we are asked to discuss verisimilitude appealing to Tarski's calculus of deductive systems: a deductive system being any set of sentences closed under the operation of logical consequence. Miller uses Cn for the consequence operation and defines a deductive system as a set of sentences A which satisfies $A = Cn(A)$. It is assumed that systems can be partially ordered by the usual relation of set-theoretic inclusion; that the set S of all sentences is the greatest, whereas the set of logical truths (L) is the smallest, and that T, the set of all truths, is complete. Miller represents the truth content of a system A as the intersection of A with T $(A \vee T)$ and, like Popper, admits that the falsity content is not a deductive system in the Tarskian sense.[29] Assuming that falsity content and truth content vary with content, Miller shows that the condition of comparability expressed in (2) is inadequate to deal with any pair of false theories. I shall formulate the gist of his argument without giving the technical details of the proofs.

Suppose that theories A and B are comparable by verisimilitude. Then, according to (2), if B is closer to the truth than A, it must either have more truth content or less falsity content. Assume that they are not identical (and that A and B are not equivalent), that $B \vee T \mathrel{\|\!\!\!-} A \vee T$; (where '$\mathrel{\|\!\!\!-}$' represents the weak derivability relation), and that B does not fall short of A in truth content. Then we only have the four following possibilities:

(a) B is true and A is true. Thus $B \mathrel{\|\!\!\!-} A$, and so $B \vdash A$ (since they are not identical). Thus $B \vee T \vdash A \vee T$. We conclude that B is closer to the truth than A is.

(b) B is true and A is false. A definitely has more falsity content than does B. Thus B is closer to the truth than A is.

(c) B is false and A is false. [In this case $B \mathrel{\|\!\!\!-} A$, so $B \vdash A$. Therefore, B exceeds A in falsity content. But if A is false, it can be proved that B has more truth content than A.] Thus B exceeds A in both truth and falsity content, so that they are not comparable.

(d) B is false and A is true. As in (c), B exceeds A in falsity content. If their truth contents are not the same, then B has more, so that no comparison is possible. Otherwise we have a subcase of (b) with A and B reversed.[30]

The upshot is that any pair of distinct theories that are comparable by verisimilitude fall under the constraints just given. If B is to be closer to the truth than A, then it must be a true theory and $A \vee T$ must follow from it. If this result is correct, Miller has shown that Popper's qualitative theory of verisimilitude (as expressed in (2)) is inadequate. In the last section of his paper, Miller takes issue with Tichý's intent to give an appropriate quantitative definition of 'verisimilitude' and discusses his toy theory of the weather. Although I want to eschew the quantitative theory, let me give briefly Miller's criticism of Tichý. Consider another rudimentary language, intertranslatable with L, that contains three atomic sentences: 'it is hot', 'it is Minnesotan' and 'it is Arizonan' symbolized by r, m and a, respectively. Now suppose that m is defined as either hot and wet or cold and dry (so that m is equivalent to $r \leftrightarrow p$), and that a is defined as either hot and windy or cold and still (so that a is equivalent to $r \leftrightarrow q$). Furthermore, p is equivalent to $r \leftrightarrow m$ and q is equivalent to $r \leftrightarrow a$. The remaining conditions in Tichý's example are unchanged. Using the new language, Smith's theory becomes $\sim r \cdot \sim m \cdot \sim a$, whereas Jones's is $\sim r \cdot m \cdot a$, (the truth comes out as $r \cdot m \cdot a$). The situation is now reversed: Jones is right in two counts and Smith in none. Miller thinks this shows that comparisons of verisimilitude cannot be as language-dependent as Tichý's counting method makes them, and that any attempt to give a suitable quantitative definition of 'truth content' or 'falsity content' is deemed to fail.

5.5 Reply to Tichý and Miller

Since Tichý's and Miller's respective results against the qualitative definition of 'verisimilitude' (although formulated differently) are equivalent, I shall offer a single reply to their objections.[31] Their argument can be restated using our old acquaintances t_1 and t_2. One may say of two false but internally consistent competing theories with comparable truth and falsity contents that t_2 is a better approximation to the truth than t_1 if, and only if, either the truth content but not the falsity content of t_2 exceeds that of t_1, or when the falsity content but not the truth content of t_1 exceeds that of t_2. More precisely, t_2 possesses more verisimilitude than t_1 if and only if t_2's falsity content is a subset of t_1's falsity content while t_2's truth content is not also a subset of t_1's truth content, or when t_1's truth content is a subset of t_2's truth content while t_1's falsity content is not also a subset of t_2's falsity content.

(8) $\quad Vs(t_2) > Vs(t_1)$ iff $(Ct_F(t_2) \subseteq Ct_F(t_1)$ and $Ct_T(t_2) \not\subseteq Ct_T(t_1))$,

or: $(Ct_T(t_1) \subseteq Ct_T(t_2)$ and $Ct_F(t_2) \not\subseteq Ct_F(t_1))$

According to Tichý and Miller, the definition in (8) (which, as far as we know, goes along the same lines as the one in (2)) cannot support attributions of higher verisimilitude to a false theory in the light of a comparison to any of its false competitors. In effect, both authors maintain that any increase in the truth content of the succeeding theory will be accompanied by a corresponding increase in the falsity content of the very same theory that counterbalances the former. Assume theory t_2 is false (though a better approximation to the truth than its rivals), then there is at least one false statement (call it r) comprising part of its content. If the truth content of t_1 is a subset of the truth content of t_2, there is also a statement (call it s) that is a member of the truth content of t_2 which exceeds that of t_1. Nonetheless, the conjunction of r and s belongs to the falsity content of t_2 but not to the falsity content of t_1. Therefore, the falsity content of t_2 could never be a subset of the falsity content of t_1 as long as the truth content of t_2 exceeds that of t_1. According to this view, increase in the truth content of a theory is always followed by increase of its falsity content. But if this is correct, all false theories are of equal status. The excess truth content of one theory over another may always be conjoined with a member of the falsity content of that theory to add proportionally to the falsity content of that theory.

Tichý and Miller proposed some amendments to (8) and developed (independently) a metric that supposedly could solve the shortcomings of the qualitative definition. However, Popper himself showed that all attempts to arrive at a metric treatment of verisimilitude are faulty.[32] For instance, they attribute equal measures of Vs to theories which should have (intuitively) different degrees of truthlikeness, in a way that contradicts the expected outcomes. Furthermore, they attribute increasing degrees of verisimilitude to false theories deduced from false predecessors, when they should have decreasing degrees of verisimilitude (given that they would have less logical content). The first response to Tichý's and Miller's objections that comes to my mind is that when we compare verisimilar theories we are concerned with the test-situations that are relevant to them and that can tell us which tests both of them have failed (that account for their falsity content) and which tests have been successfully passed by the better one but not by its rival. In any case, we should not be worried for the results of logical operations that can be performed in one theory but not in the other.

It is surprising, to say the least, that neither Popper nor any other commentator seems to give proper consideration to the caveat in (2*).[33] But if we are to exclude tautologies from the truth content of an empirical theory, I see no reason for not being at least cautious with tautological moves. If tautologies are not welcomed because they do not have empirical content, logical operations in general, since they yield statements that do not usually add to the empirical content of their antecedent component, should also arouse suspicion. I think that if we make some further restrictions to (2*) in such a way that we either block or limit the use of logical operations as a legitimate way to add to the truth-content class of a theory, then we may respond successfully to the objections raised above against the qualitative definition of 'verisimilitude'. Let me present two ways of achieving this end.

According to Popper, theories are sets of statements over a particular language. Thus, one might consider that a theory is a logically complex statement. If we think about what Popper wants to convey with the notion of a theory's truth content being contained in another theory's truth content, it seems that this notion is not fully captured by the presupposition of Tichý's and Miller's objection. The idea is that if $Ct_T(t_1) \subseteq Ct_T(t_2)$ and t_1 and t_2 share some test-situations, the same true consequences will generate when the same logical operations are applied to t_2's responses to those tests. Let us see if we can cash that out by developing Popper's caveat concerning tautologies and my suggestion on tautological operations. How, then, can we preclude the logical operations that generate the problem?

To account properly for the excess truth content, and the response of a theory to a test-situation, we must focus in the information that a theory actually gives about the world. Let us assume that theories are stated in an ideal language L, which meets the following conditions: (i) the predicates of L are as analysed as they can be (so that the concepts they express can be grasped without further analysis); (ii) L is consistent; (iii) L is closed under the operation of logical consequence. If we think of theories as formulated in this canonical language (in which all the predicates of the language are as simple as possible) we can compare directly the cardinality of the predicates. We need a function from the truth content to a subclass of the things that are actually said about the world by a theory; that is, a function of the truth content to the true atomic sentences of Ct_T or the negated statements that are true. To this effect, we want to single out the smallest set of atomic sentences from which the other sentences of the truth content of the theory are deducible, with the caveat that no conjunction will be allowed. In addition, we will characterize this subset as the cardinality of its class of atomic statements. Once we have the subsets for both t_1 and t_2, we will attribute greater Vs to t_2 according to the formulation in (9). (Let '$|A|$' stand for the cardinality of A, and '$AT(Ct_x(t_y))$' stand for the set of atomic sentences of the x-content of theory y):

(9) $Vs(t_2) > Vs(t_1)$ iff

$((|AT(Ct_T(t_2))| > |AT(Ct_T(t_1))|) \,\&\, (|AT(Ct_F(t_2))| \leq |AT(Ct_F(t_1))|))$

or $(|AT(Ct_T(t_1))| \leq |AT(Ct_T(t_2))|) \,\&\, (|AT(Ct_F(t_2))| < |AT(Ct_F(t_1))|))$

This provides a simple and elegant solution to the difficulty at issue. By requiring that all true statements which count as legitimate additions to the truth-content subclass of a theory be atomic ones, we are blocking the sort of objection examined in the foregoing. By extending the same requirement to

Verisimilitude 139

the falsity-content subclass one might dissolve the criticisms of Tichý's and Miller's to (8) and (2). In effect, many logical operations, in so far as they yield molecular statements, would be blocked by (9). In particular, conjunctions and disjunctions should not represent any problem for the amended qualitative definition of 'verisimilitude', since their conclusions will not qualify to add to the truth content or the falsity content of any empirical theory. At least three questions are in order at this moment. The more general question will be: is it desirable (and even possible) to ban all logical operations? The next question is: what do we do with the logical operations that yield atomic statements as a result? (For example, those that merely unpack statements or validly separate any of their components.) What about logical operations that simply transform a statement into an equivalent?

These questions suggest that (9) is far from being a straightforward and complete solution. It is doubtful that the answer to the general question could be affirmative, though one may respond by saying that the intention is not to rule out logical operations but only to preclude their conclusions (inasmuch as they are molecular statements) from entering the truth-content subclass. However, the remaining questions show that (9) by itself is insufficient to produce the result that we want. My position, however, is not hopeless. Another amendment to (2*) can do the trick.

To introduce my second suggestion let me define the notion of Empirical Content (sometimes called also *informative content*). According to Popper, the empirical content of a statement of a theory (Ct_E) is the class of all observational statements (basic statements) which *contradict* the theory. In the terminology of *LScD* it is nothing else than the class of potential falsifiers. It is important to mention that Popper excludes from the class of basic statements (from which statements that play the role of potential falsifiers come) the negations of almost all basic statements as well as disjunctions and conditionals obtained with them. The reason he adduces for this exclusion is this:

> [W]e do not wish to admit conditional statements such as 'If there is a raven in this room then it is black', or 'If there is a mosquito in this room then it is an *anopheles*'. These are no doubt empirical statements; but they are not of the character of *test-statements* of theories but rather of *instantiation statements*, and therefore less interesting, and less 'basic', from the point of view of the theory of knowledge here expounded; a theory of knowledge which holds the empirical basis of all theories to be tests; or in other words, attempted refutations.[34]

In a parallel way I want to exclude from the truth-content subclass not only tautologies (as does Popper) but also the results of some logical operations. In particular, I want to exclude the results of all logical operations that yield conclusions with lesser empirical content than the statements they come from, and logical operations that merely transform statements into their equivalents, as well as the addition of trivial truths. This exclusion is congenial to a fallibilist epistemology in which increase in empirical content is a desirable aim of science. So, broadly speaking, on my view an increase in truth content should also give us an increase in empirical content, if it is to count to support an increase in verisimilitude. Let me state, then, the new requirement: one should count as an addition to the truth content of a theory only nontautological, true statements which have at least as much empirical content as the statements that already account for the empirical content of a theory, provided they are not mere reiterations or simple transformations of the statements from which they are deduced nor trivially true. We might see this more easily in the next formulation. Let $\boldsymbol{Ct}_{T*}(t_2)$ be the excess truth content of t_2 over t_1 (in other words $\boldsymbol{Ct}_T(t_2) - \boldsymbol{Ct}_T(t_1) = \boldsymbol{Ct}_{T*}(t_2)$); and $Ct_{T*}(t_2) = \{x : x \in Ct_T(t_2) \ \& \ \sim x \in (Ct_T(t_2) \cap Ct_T(t_1))\}$, then

(10) $Vs(t_2) > Vs(t_1)$ iff $\boldsymbol{Ct}_T(t_2) > \boldsymbol{Ct}_T(t_1)$ and $\boldsymbol{Ct}_F(t_2) \leq \boldsymbol{Ct}_F(t_1)$,

or: $\boldsymbol{Ct}_T(t_1) \leq \boldsymbol{Ct}_T(t_2)$ and $\boldsymbol{Ct}_F(t_2) < \boldsymbol{Ct}_F(t_1)$; and

(y) (x) $((y \in Ct_{T*}(t_2) \ \& \ x \in Ct_T(t_1)) \rightarrow ((\boldsymbol{Ct}_E(y) \geq \boldsymbol{Ct}_E(x))$

$\& \sim (y \leftrightarrow x)))$

However, this might not be sufficient yet. For suppose statement p belongs to the empirical content of t_2. Then, the disjunction of p with any false statement (assume one that comes from the falsity content of t_1) is a true statement, is not a tautology, is not equivalent to p, is a logical consequence of p, and it might have more empirical content than p, satisfying all of my requirements in (10). One could give a list of undesirable logical operations (à la Popper) that should be excluded, but perhaps there is a better solution. Let us consider a combination of the amendments in (9) and (10). This will give us:

(11) $Vs(t_2) > Vs(t_1)$ iff

$(((|AT(Ct_T(t_2))| > |AT(Ct_T(t_1))|) \ \& \ (|AT(Ct_F(t_2))| \leq |AT(Ct_F(t_1))|)$

or $(|AT(Ct_T(t_1))| \leq |AT(Ct_T(t_2))|) \ \& \ (|AT(Ct_F(t_2))| < |AT(Ct_F(t_1))|))$

And (y)(x)$((y \in Ct_{T*}(t_2) \ \& \ x \in Ct_T(t_1)) \rightarrow ((\boldsymbol{Ct}_E(y) \geq \boldsymbol{Ct}_E(x))$

$\& \sim (y \leftrightarrow x))$

It appears that this will do it. It excludes undesirable logical operations (such as conjunction and addition) that yield molecular statements; it is consistent with the desideratum of searching for theories that are more informative (empirically speaking) than their discarded predecessors; and it precludes statements which really add nothing to a theory's truth content from being counted as its assets. In addition, it is easy to see that (11) is consistent with the amendment in (2*) and that it meets all desiderata in (4). Popper values highly a treatment of verisimilitude that is consistent with the general lines of his epistemology. On such lines, logically weaker statements are, individually considered, less verisimilar than logically stronger ones, and when incorporated to sets (such as the subset of true statements) add little to the logical strength of the whole class. For these reasons, Popper also commends theories with high degree of logical strength (high logical improbability) over their weaker counterparts. However, a problem remains to be solved. In the original formulation of verisimilitude, Popper tried to establish a parallel between the Tarskian idea of truth and his formulation of verisimilitude. He pointed out the need for a treatment that would give to this notion the same objective character and the same regulative character that Tarski gave to his idea of objective or absolute *truth*.[35] In '*NV*', Popper stressed the need of finding a satisfactory definition of this notion that could produce 'a measure of verisimilitude which like truth would be invariant with respect to translations into other languages'.[36] Unfortunately, since both (9) and (11) rely on a syntactical criterion, they could be considered as language-dependent and unable to comply with this feature. However, solving this problem does not fall within the scope of my objectives in this book, so I shall defer this task for another occasion.[37]

5.6 The quantitative measure of verisimilitude

As already suggested, there are reasons to think that no metric yields desirable results for a quantitative measure of verisimilitude (henceforth QVs), and certainly, Miller contends that any attempt along the lines of Tichý's rudimentary theory is hopeless. Without the intention of proposing such a metric myself, let me give a simple way of making sense of the notions involved in ascriptions of QVs that works perfectly well for toy theories. It seems that a construal in terms of the arithmetic ratio between the cardinality of the set of statements that comprises the theory and the cardinality of its true statements suffices to this effect. Suppose that our theory, modelled after Tichý's, is comprised of the following three statements:

(1) it is raining (r)

(2) it is hot (t)

(3) it is windy (w)

Call this theory the 'weather' theory for convenience. Let S be the cardinality of the set of statements of the theory, t_F the cardinality of the subset of false statements of the theory and t_T the cardinality of the subset of true statements of the theory. We can easily determine the ratio of true statements with regard to S, the whole set of statements that comprises the theory.[38] For the six (basic) disjunctive normal forms of our simple theory, we would obtain the following ratios, under the assumption that the state of affairs makes true r & t & w:

(a)   r & t & w $t_T/S = 3/3 = 1$

(b)   r & t & ~w $t_T/S = 2/3 = 0.66$

(c)   r & ~t & w $t_T/S = 2/3 = 0.66$

(d)   r & ~t & ~w $t_T/S = 1/3 = 0.33$

(e)   ~r & ~t & w $t_T/S = 1/3 = 0.33$

(f)   ~r & ~t & ~w $t_T/S = 0/3 = 0$

Call this the 'ratio technique' to estimate QVs. In the extreme cases (no true statements or no false statements), the corresponding ratios yield non-controversial appraisals of plain falsity or absolute truth.[39] Assuming that the theory we are assessing has at least one true statement, the elementary formula t_T/S will give us a (positive) estimate of QVs and the elementary formula t_F/S will give us an estimate of its relative falsity (if the theory has at least one false statement). For the case in which there are no true statements in the theory we have to say that the truth-content subclass is empty and assign a zero degree of Vs to the theory, which accords with Popper's intention and with our intuitions.[40]

If, as stated above, it is raining, hot and windy, the weather theory is absolutely true, and it should receive the maximal measure of QVs. Accordingly, its distance to the truth is zero (because it coincides with it). False weather-related competing theories that are wrong in two counts or one count will have, respectively, values of QVs of 2/3 and 1/3, which accord to our intuitions; and a theory that is wrong in all three counts will have a value of QVs of 0/3 which is the same as zero (i.e. it will be absolutely false, also in accord with our intuitions). It is easy to see, also, that false weather-related competing theories receive a better mark for QVs as a function of having more true statements (and presumably a higher truth content). This technique yields satisfactory results for any finite group of related (but competing) theories that differ only in the respective number of true and false statements (the number of variants being also restricted by the number of statements in the theory).[41] In addition, it calls for a different topological representation of the possible variants.

To see the peculiarities of this representation, let us adopt the following convention. Assume that we can put theories along a horizontal continuum in accordance with their degree of *Vs* (as measured by the ratio technique). We can say that the degree of *Vs* of a theory (when the appraisal can be made) lies between 0 and 1, with the exclusion of both extremes.[42] Let the maximum degree of *Vs* equal one. But maximum degree of *Vs* is just plain truth. So let us attribute this measure only to (absolutely) true theories, that is theories which are not (strictly speaking) verisimilar but true.[43] Let complete lack of *Vs* equal zero. Then zero degree of verisimilitude equals absolute falsehood (instead of tautologousness, as Popper holds). On the other hand, though this technique supports the attribution of a certain measure of *Vs* to an isolated theory, it is not clear how to draw any epistemological conclusion out of a single value. However, this problem can be remedied when we consider at least two related theories. As a comparative notion (as I also take the notion of falsifiability to be), *Vs* only makes sense when it is decided in the light of at least one competitor. I think this point, which has been neglected in the literature, is crucial to comprehend the way the criterion works. We can see that the previous rules to rank-order theories conform to the principles of Popper's epistemology by taking into account that (i) absolutely true theories, that is, theories with an empty falsity-content subclass cannot, properly speaking, be called verisimilar: they are *true* in the full extent of the word; (ii) tautological theories are excluded, not because they cannot have a nonempty truth-content subclass (or because they have an empty falsity-content subclass) but because they are not *testable* and self-contradictory theories are excluded because they do not meet the basic condition of empirical science. All other theories should be accorded various degrees of *Vs* and, therefore, can be properly represented as occupying different positions in the continuum. Since any comparison of *Vs* is a comparison of the size of the respective truth-content subclass and the falsity-content subclass of competing theories (a comparison of their *ratios*, according to the technique just described), the continuum admits of an indefinite number of proper fractions between 0 and 1, with the exclusion of both extremes. This feature is desirable since any comparison involves a finite number of theories, perhaps two and almost certainly just a small handful of them. (That the number of possible theories is infinite does not pose any problem, since there is no actual situation in which one may be confronted with an infinite number of theories and needs to select among them. However, if the objector insists, one could rephrase the definition to accommodate an infinite number of values, for nothing important hinges on this issue.) Accordingly, we can represent degrees of Vs as follows:

$$(12) \quad 0 \xrightarrow{\quad Vs_1 \quad\quad Vs_2 \quad\quad Vs_3 \quad\quad Vs_4 \quad\quad Vs_5 \quad} \xleftarrow{\quad} 1$$

Absolutely false theories[44]   Absolutely true theories

Perhaps I need to re-emphasize the motivation for one of the restrictions proposed in (11). Precluding the use of logically inferred statements (by using tautological moves) to build the truth-content subclass conforms better to a workable definition of 'verisimilitude' for the following reasons. Failure to do so makes it practically impossible to differentiate between the truth content of any two arbitrary true propositions, since any proposition implies tautologically an infinite number of statements. As we know, truth is transmitted from premises to conclusion, then if a statement is true its logical consequences must be true.[45] But, in empirical science we are not interested in increasing the number of true consequences *simpliciter*. We want true consequences that have more empirical content than the premises from which they were inferred. Thus, if we restrict the consequences of a given statement (theory) to statements that are not obtained tautologically, we can say that either we obtain true consequences that might add to its truth content or we may obtain false consequences that increase the statement's falsity content. In either case, we would obtain statements that can be used to determine the status of a theory with respect to its truth content, its falsity content and accordingly its degree of Vs. On the other hand, if a theory has *only* false statements, I see no reason to accord to it any (positive) degree of Vs just because many true statements follow from it by conditionalization (which precludes its truth-content subclass from being empty). It seems preferable to accord to this theory zero degree of Vs to indicate maximal distance to the truth *within* the family of theories to which it belongs. This situation should not be confused with the case in which we have two false related theories (with non-empty truth-content subclasses, according to the restrictions of (11)). In such a case, a comparison between them in terms of truthlikeness is meaningful only if the two theories besides being false (presumably falsified, though not absolutely false) minimally meet the two following criteria: (i) they have been corroborated at least once; and (ii) they have some residual empirical content. But, again, they both cannot be absolutely false or else the comparison would not be meaningful.[46]

Unfortunately, the ratio technique to estimate degrees of QVs has several shortcomings that render it useless. Firstly, it works only for very simple theories formulated by means of a finite set of statements, which cardinality we know. But, as is currently accepted, the number of statements of any epistemically interesting empirical theory is unlimited and probably infinite (not to mention that what makes a theory strong lies precisely in its unknown true consequences). This means, briefly stated, that this way of calculating degrees of QVs is inapplicable to the sort of theories in which the epistemologist is interested.[47] Secondly, this technique ignores the basic definition of Vs according to which that notion is tied to the logical consequence operation for both the truth-content and falsity-content subclasses. Disregarding tautologies (and many statements obtained by using tautological, that is, truth-preserving moves, as indicated above) one can still expect that the number of true consequences of a theory is relatively large; in any case that is larger than the number of its component statements and its disjunctive normal forms, and this fact alone seems to defeat the ratio technique and other procedures of similar persuasion used to estimate degrees of QVs. If we tried to use Popper's scarce examples as a model for the sort of statements that qualify as legitimate members of a given truth-content subclass and attempt to draw the corresponding true consequences for the weather theory, we would obtain statements such as: 'it is not still'; 'people who go out under the present conditions normally use umbrellas'; 'flags in flagpoles are wavy'; 'the thermometer reads on or over 85F', and many others of that ilk.

What sort of predictions does this theory make? Hypothetical statements such as 'if you go out without an umbrella you will get wet'; 'if that tree has loose limbs (or leaves) then they will fall', etc. It is hard to think of a false prediction that can be made by using the (absolutely) true version of this theory, provided we restrict ourselves to the set of statements given and stick to the conditions stipulated in (11). But if we use one of the weaker false theories, then we can make false predictions such as: 'the thermometer will give a reading which is less than 85°F' (assuming we have previously stipulated that any reading of temperature less than 85°F does not count as hot), as well as some true predictions which will increase the number of its true consequences (by the way, consider that theories a–e above can share some of its respective true consequences). The complexities and difficulties posited by a simple theory such as this illustrate how it could be the case for a real theory that is comprised of many statements (both true and false).[48] However, many of the difficulties can be solved (or eschewed in favour of an intuitive approach) if we (a) restrict the discussion to the case of competing theories and (b) qualify the theories' statements in such a way that we are able to rank order them by epistemic value. Let me attempt to sketch this proposal.

Assume that two distinct theories are genuine rivals if (i) they speak about the same sector of the world (ii) they have passed at least one similar test and (iii) they conflict at least in one aspect that involves description and prediction of facts in the sector of reality that the theories intend to explain and represent. Let me introduce roughly a way to rank the epistemic value (Ev) of the different statements (with empirical content) that comprise the theory. Let us assume that descriptive true statements have, regularly, less Ev; that explanatory statements have medium Ev; that repeatedly corroborated hypotheses count as descriptive true statements; that falsified hypotheses count as false statements (hence lack Ev); and that surviving a severe test that the previous theory could not survive has (temporarily) the highest Ev. Then, we can use the notion of Ev to substantiate attributions of Vs as well as our choice between competing theories.

The simplest ideal case is that in which theories t_1 and t_2 are genuine rivals and the first has failed test W, whereas the second has survived the same test and has both more truth content and less (or equal) falsity content than its rival.[49] Of course, we also expect that t_2 has successfully passed all relevant empirical tests that its rival has passed; or, in other words, that before test W, the respective degrees of corroboration of t_1 and t_2 are the same. There is no doubt that in this case: we have good reasons to prefer t_2 over t_1 and to consider t_2 as a better approximation to the truth than its rival because t_2 exceeds t_1 in corroboration and Vs while having more Ev regarding the pertinent statements. To see why, presume that all theories are false. Then we will attribute a higher degree of Vs to a theory if it makes the same predictions, it survives at least one of the tests that the other failed and it has more Ev. To make things easier, suppose that t_1 stands for Ptolemy's theory and t_2 stands for Copernicus's. Then we know that concerning the problem of the astronomic positions for the planets, similar and perhaps equivalent predictions can be drawn from both t_1 and t_2. Likewise, with better techniques of aided-eye observation, both theories can yield equally good values to be used in astronomical tables. But Ptolemy's theory cannot accommodate the phases of Venus, and, so to speak, fails this test. Copernicus's theory, by contrast, can accommodate this fact and survive this test obtaining more Ev. Consequently, we can assign a higher degree of verisimilitude to Copernicus's theory since it meets our criteria above. But this is perhaps only the ideal situation while other cases of theory comparison may be more challenging.

Suppose that theories t_1 and t_2 are genuine rivals, that the first has failed test W and the second test Z, but are equally good in every other respect. Since I intend to give a characterization of false theories that have (presumably) a different degree of Vs, improving our intuitive assessment with the notion of Ev, we would need to compare W and Z in order to substantiate any appraisal. Only if the respective degrees of severity of W and Z are different can we select a theory. Otherwise, we would need to consider the compared theories as equally good until a common test is found. Call this sketch 'the intuitive approach to Vs'. Obviously, the intuitive approach to Vs is liable to several criticisms. One may object, for example, that it is too elementary to be useful. It makes comparisons of theories by Vs a child's game. Moreover, it could be argued that the notion of Ev is too poorly defined to be of any use. Another

Verisimilitude

objection may be that the approach would not be useful for theories that have not failed any test so far, either because they (or one of them) have not been tested yet, or because they have been tested and have survived the same tests.[50] In the first case, the answer suggests itself: two competing theories that have not been tested cannot be accorded a positive degree of corroboration and, correspondingly, cannot be accorded a positive degree of falsity-content. In this case, both theories, *qua* conjectures, are equally good/bad, and we lack grounds to choose between them. The upshot is that a comparison in terms of Vs necessitates that the theories be previously tested. The second case, in which both theories have survived the same tests looks also to be a case in which there is no criterion to choose between them, unless the situation is such that theory t_2 has passed the same tests as theory t_1, plus more because the latter has not been subjected (for whatever reason) to some of the tests yet. In such a case, since ascriptions of Vs are not definitive, one could choose tentatively the theory that has passed more tests, on the grounds that it is the best corroborated, and it may be also the one closest to the truth[51].

Notes

1. The substance of the theory of verisimilitude is to rank theories according to how well they approach the regulative idea of truth.
2. This statement requires some qualification. With the help of the concepts discussed in the previous chapters, we can tell how a good theory looks (it would be well-corroborated and falsifiable in a high degree) even before the theory is tested. We could say also which ideal theory would be even better, given certain conditions (e.g. its potentially passing some other tests). And this meta-knowledge supports a vague idea of progress in terms of what Popper calls 'relative potential satisfactoriness'. However, we are interested now in a more straightforward and more comprehensively formulated criterion of progress. Since truth does not manifest itself, we need to know when we are in the right track and be able to decide whether we should stay on it. See *CR*, p. 217; and 'Rsr', pp. 93ff.
3. This assertion is controversial. I do not want to suggest that Popper is committed to it, since he allows that a theory may be true (though perhaps we might never discover it). I deal with this problem at length later.
4. The way Tarski formulates his theory does not make use of the notion of truth having to do with what the world is like, though he provides an articulated way of understanding truth as correspondence. Perhaps Popper is not giving a faithful account of Tarski's theory of truth, but, given our purposes here, we need not worry about this issue.
5. Cf. *OS*, p. 46. For a detailed treatment of Popper's appreciation of Tarski's theory of truth see Miller 1999.

6. *CR*, p. 225. For Popper's account of the epistemological implications of Tarski's objective theory of truth see: *OS*, Vol. 1, p. 144 and Vol. 2, p. 369; *UQ*, p. 130; *RAS*, pp. 73, 79, 266, 273–5; *MF*, pp. 174–5. For a more technical treatment of the definition of the notion of 'fulfilment' by appealing to finite sequences, see Popper 1955.
7. *CR*, p. 226.
8. Although this is not a consequence class in the Tarskian sense, Popper offers some reasons to treat it as such. (See *CR*, pp. 393–7; *OK*, pp. 48–52.) In *CR* he wrote 'truth-content' and 'falsity-content' hyphenated. In *OK* he restricted the use of the hyphens to adjectival contexts. I will apply that convention here.
9. *CR*, p. 234. There is some ambiguity in Popper's discussion of this idea since he uses the same notation to refer both to statements and to theories; to measures of content and to the classes of statements that constitute those contents. To eliminate the second ambiguity, hereafter I use (as in (1) above) boldface italic print to denote *measures*, and ordinary italic to denote *classes of statements*. As far as the first type of ambiguity is concerned, given that theories can be considered as systems of statements one may think that the measure of verisimilitude simply carries over from statements to theories. However, although this does not seem to be the case, for the purposes of examining the notion of distance to the truth, whatever is said about statements can be safely applied to the case of theories with no risk of confusion. It is worth mentioning, in passing, that Popper commits the same ambiguity in all places in which he discusses verisimilitude.
10. Strictly speaking, we can make progress towards the truth just by adding to a theory's truth content provided we keep its falsity content unchanged.
11. 'Verisimilitude is so defined that maximum verisimilitude would be achieved only by a theory which is not only true, but completely comprehensively true: if it corresponds to *all* facts, as it were, and of course, only to *real* facts. This is of course a much more remote and unattainable ideal than a mere correspondence with *some* facts (as in, say, "Snow is usually white")' (*CR*, p. 234). By contrast, logic and mathematics are fields in which we can reach absolutely certain truth, but this empirically empty truth does not say anything about the world. See *OS*, Vol. 2, p. 13.
12. Let me stress that (1) gives us a definition of the notion of 'distance to the truth' for statements. Consequently (2) should be understood as using the very same notion for the case of theories. On the other hand, to perform a comparison of the corresponding degrees of verisimilitude between competing theories, a further constraint needs to be introduced. We need to deal with genuine rivals, that is, theories which talk about the same sector of the world.
13. Popper says explicitly that in this passage he is using the word 'probability' in the sense of the calculus of probability. However, a few paragraphs later he slips in the notion of absolute logical probability, which he takes to be required to mark a fundamental difference between laws of nature and laws of logic. On his view, laws of logic have both maximum logical probability and nill empirical content, exactly as the first Wittgenstein held. Wittgenstein introduced the notion of probability in *Tractatus* (5.15) by saying that the probability that one proposition gives

to another is the ratio of the number of their common truth-grounds to the number of the truth-grounds of the first, where a 'truth-ground' means a distribution of the truth-values to the truth-arguments of the proposition needed to verify the proposition. According to this, a tautology has probability 1; a self-contradiction probability 0; and any other elementary proposition gives to each other the probability $\frac{1}{2}$. Popper sometimes adopts this course, as we can see in *LScD*, p. 318 where he defines the absolute (logical) probability of x as: $p(x) = p(x, \sim(x \& \sim x))$ and concludes that the absolute logical probability of any supposed law of nature is less than any assignable fraction. Popper does not agree with Wittgenstein's idea that laws of nature are propositions of accidental universality, nor does he accept the assumption of logical atomism that each elementary proposition gives to each other the probability of $\frac{1}{2}$, but he concurs with Wittgenstein that necessity belongs only to tautologies.

14. *CR*, pp. 393–4. All these *desiderata* spring from (1) and (3) above, and they contribute to what has been called the *logical* (qualitative) definition of 'verisimilitude'. Popper also offers a probabilistic definition of 'verisimilitude' (sometimes called *quantitative*). In this chapter, I shall focus mainly on the former. Since according to (1) $(\forall t)\ \boldsymbol{Ct}_T(t) \neq 0$, (v) is vacuously true unless $(\exists t)\ \boldsymbol{Ct}_T(t) = 0$ which does not seem to be the case at that stage.

15. Cf. ibid., p. 396. In the third expression a remains ambiguous and it can be replaced by a false or a true statement, thus obtaining different values of *Vs*. Later on, I would suggest dropping any assignments of *Vs* to self-contradictions.

16. *OK*, p. 48.

17. The falsity-content subclass is not a Tarskian consequence class (and it is not always the case that the truth-content subclass is a Tarskian consequence class). But Popper offers a relativized definition in the following terms: $Ct_F(a) = Ct(a, A_T)$ (where A_T is the intersection of A, the content of a, with T, the Tarskian system of true statements) which means that the measure of a's falsity content is 'the (measure of) the relative content of a, given the truth content of A_T of a; or, in still other words, the degree to which a goes *beyond* those statements which (a) follow from a and (b) which are true' (*OK*, p. 52).

18. Popper seems to see this problem in a different way. In his opinion, true theories can have different degrees of verisimilitude according to their respective empirical content or their respective logical strength (i.e. true theory t_2 has a greater degree of verisimilitude than true theory t_1 if t_2 has more empirical content – or if it is logically stronger – than t_1). Although I have some trouble understanding this doctrine, I believe what Popper has in mind is something like this: the aim of science is to grasp the absolute truth. A true theory with higher empirical content has more to tell about the whole truth than a less informative competitor, so even if they are both true, the former is closer to the whole truth and hence more verisimilar. But we can perform an equivalent comparison solely by means of the notion of empirical content and avoid this difficulty. After all, the real value of the notion of verisimilitude is supposed to lie in the treatment of unequally false competitors. For Popper's view see Popper 1976a, pp. 153–5.

19. Suppose it is now 9.45 p.m. Then we can say that the statement 'It is now *between* 9.45 p.m. and 9.48 p.m.' is closer to the truth than the statement 'it is now between 9.40 p.m. and 9.48 p.m.' (cf. *OK*, p. 36). Note, however, that this procedure can be applied only to a very narrow class of statements.
20. *CR*, p. 235.
21. At least as good as its better-falsified competitor. On the other hand, we have reasons to expect that any theory will eventually fail a genuinely severe test (remember the testability requirement) and for this reason alone there is not a significant difference between failing it at the beginning or a long time ahead along the road (cf. *CR*, pp. 242–3).
22. There is an indisputable relationship between verisimilitude and corroboration: namely that a corroborated hypothesis may indicate that we have hit on the truth and, consequently, it may count for the truth content of the theory to which it belongs, but this is not the kind of relation that legitimizes any attempt to define the former notion in terms of the latter – among many reasons because the notion of truth is timeless, whereas appraisals of corroboration are always indexed to a point in time and a set of accepted test-statements. On the other hand, discussing verisimilitude in terms of corroboration is a great temptation for many commentators. One who succumbs to it is John Harris, who makes a mistake reminiscent of Lakatos's. Harris thinks that corroboration is a special case of the problem of verisimilitude and treats it as a function of the agreement of a theory with the experimental data. But the inductive spirit of such a proposal makes it completely unacceptable within a falsificationist framework (cf. Harris 1974). On the other hand, there are creative uses of this relationship. One such use can be found in Agassi's suggestion to save verisimilitude by reconciling the view that progress in science is *empirical success* with the view that progress in science is *increase in verisimilitude*. Agassi purports to accomplish this by defining an increase in verisimilitude as the combination of an increase in truth content and a decrease in falsity content, and by urging the point that 'a theory is more verisimilar than its predecessor if and only if all crucial evidence concerning the two goes its way' (cf. Agassi 1981, p. 578).
23. Tichý 1974, p. 156. Note that this definition is supposed to restate (2), hence any criticism to it affects (2). It is not altogether clear that Tichý captured correctly what Popper says, since he is using A_T to refer both to a measure and to a class and there is no straightforward connection between the truth content defined as a *measure* and the same notion defined as a *class*. This ambiguity, which is omnipresent in secondary literature, seems to be licensed by Popper's alternative definition of 'verisimilitude' in terms of the subclass relation (see *OK*). In what follows, I am correcting some obvious notational mistakes in Tichý's argument.
24. Cf. ibid., pp. 157–60. Tichý gives Popper's probabilistic definition along these lines: take A and B as theories, let $p(A)$ be the logical probability of A and $p(A, B)$ the relative logical probability of A given B. Then the truth content of A is $1 - p(A_T)$, and its falsity content is $1 - p(A, A_T)$. Verisimilitude can then be expressed in terms of ct_T and ct_F in two alternative ways: (i) $Vs(A) = ct_T(A) - ct_F(A)$ or (ii) $Vs(A) = (ct_T(A) - ct_F(A))/(2 - ct_T(A) - ct_F(A))$.

25. Ibid., p. 158. Tichý gives a table with the pertinent values (vs_1 and vs_2 stand for the two previous probabilistic definitions of verisimilitude) for three false theories of L:

	$\sim p \cdot \sim q$	$p \cdot q \cdot \sim r$	$\sim p \cdot \sim q \cdot \sim r$
CT_T	5/8	6/8	6/8
CT_F	1/3	1/2	1/2
Vs_1	7/8	2/8	2/8
vs_2	21/25	1/3	1/3

26. See note 25 for values.
27. Ibid., p. 159. More clearly, Tichý requests to put the theory in disjunctive normal form, count the negation signs and divide that by the number of the (conjunctive) constituents of the disjunction.
28. Thus one way in which a theory can grow towards encompassing the whole truth is simply through an increase in content. But, it can obviously be objected, and rightly, mere aggregation of consequences is of no use in itself; we want these new consequences to be *true*. After all, any old fairy-tale, however preposterous, if grafted on to Galileo's theory, will beget new consequences. But we would be disinclined to accept such an augmented theory, in contrast to Newton's, as a step in the direction of the truth. (Miller 1974, p. 167)
29. Cf. ibid., pp.168–9.
30. Ibid., p. 172.
31. I should mention that, after twenty years, Miller still considers right his strictures of the qualitative notion of verisimilitude. For obvious reasons, I disagree with this opinion as well as with his dismissal of verisimilitude as the appropriate aim of science, in favour of truth (though he might be more faithful to the spirit of Popper's theory on this score). Acknowledging the great importance of Miller's contributions to the theory of verisimilitude and Popper's acceptance of the tenability of his criticisms, I prefer to look in another direction (which I will make clear in due course). On the other hand, Miller offers an algebraic theory of distance to the truth that is supposedly an improvement on Popper's 'refuted' views in this matter and is not liable to the language-dependence objection because he identifies intertranslatable propositions so that their degree of truthlikeness and the comparisons by verisimilitude are fully independent of the language in which the propositions are formulated. Given its technicality I will not discuss it here. (Whether Miller's theory is correct I cannot tell.) The interested reader should see Miller 1994, pp. 202–17.
32. See: Popper 1976a. I will eschew all reference to Popper's technical proposal to solve the shortcomings of his (2*). On the other hand, since in this paper he still considers (2*) correct, and my own suggestion applies equally well to (2*) or (8) I can safely ignore what is not crucial for my argument.

33. Tichý mentions the point in passing, but he considers that as a slip since other statements in the same page in which Popper introduces the restriction seem to be in conflict with it. In any case, he believes that the difference is just 'marginal'. Cf. Tichý 1974, p. 156, note 1.
34. *CR*, pp. 386–7.
35. Verisimilitude, I hope is clear by now, is not a function of the available evidence, although (like truth itself) may be a function of the truth-value of the evidence.
36. Popper 1976a, p. 147.
37. Since the class of statements that constitute a theory can be infinite, one may limit it by admitting into the universe of statements only those which one conjectures to be relevant to the problem situation in hand. Such restriction may furnish a solution to the language-dependency objection.
38. Of course, by the same token, we can also determine the ratio of false statements of the theory with regard to S. If we considered the theory as a finite set of statements and treated these as primitives allowing only conjunction between atomic sentences (i.e. under the assumption that the theory can only be developed through four basically distinct conjunctions) then we would obtain symmetric ratios for t_T/S and t_F/S. However, to avoid unnecessary complexities, I want to stay away from the latter, and restrict my suggestions to the former. To justify this move, I can appeal to Popper's warning on the nature of falsity content. He calls that a 'content' only for 'courtesy' since it is not a consequence class in the Tarskian sense.
39. Unfortunately, we arrive at different results if we rank-order a–f by using the notion of distance to the truth (D_T), as expressed by the formula: 'truth content – falsity content'. Assuming that these contents can be measured by t_T and t_F respectively (in the sense described above), and replacing t_T/S by the appropriate fraction, the ratio yields:

(a) r & t & w $t_T - t_F = 3 - 0 = 3; D_T/S = 3/3 = 1$

(b) r & t & ~w $t_T - t_F = 2 - 1 = 1; D_T/S = 1/3 = 0.33$

(d) r & ~t & ~w $t_T - t_F = 1 - 2 = -1; D_T/S = -1/3 = -0.33$

(f) ~r & -t & ~w $t_T - t_T = 0 - 3 = -3; D_T/S = -3/3 = -1$

Now we obtain negative values, and for the absolutely false (f), the value Popper suggests should be accorded to contradictions.
40. If the theory has no true statements whatsoever, it does not require much argumentation to say that it is false. In this case it does not make sense to talk about its 'distance to the truth', unless one uses this notion as a limiting reference.
41. I am relativizing the notion of QVs to families of theories (FT) (thus, $Ct_T(t_x, FT)$, etc.). Another problem of the ratio technique is that it yields counterintuitive results when the calculations of QVs for two theories (t_1 and t_2) are performed independently. For example, let $p_1 \ldots p_n$ be the set of statements of a theory; '~' an indication of falsity, and suppose t_1: {p_1, ~p_2, p_3}, whereas t_2: {p_1, ~p_2, p_3, ~p_4, p_5, p_6, ~p_7, p_8, p_9}; then if we measure QVs in the way just

indicated $QVs(t_1) = QVs(t_2)$. But obviously t_2 is more informative than t_1 and should have greater Vs on any measure. To avoid this undesirable result we will use as the denominator the cardinality of statements ($S+$) in the theory that has more statements within the family with which we are working. So $QVs(FT) = t_n/S+$, which gives different values for $QVs(t_1)$ and $QVs(t_2)$. This proposal can be generalized to the whole family of theories, in which case one of them would be the comprehensively true theory of the world. The ratio technique would run into trouble when we consider a theory with an infinite number of logical consequences. But there is reason to think that the required comparisons can be extended to cases in which we have an infinite set of statements, because any set (even an infinite $S+$) can be mapped onto a line with 0 and 1 in the extremes. Since there are many ways to accomplish this task, the challenge would be to find the algorithm that yields the right measure for our purposes.

42. It should be obvious that this is an extension of a similar representation devised by Popper to depict graphically the degree of falsifiability of diverse theories. On the other hand, I think we are entitled to use the same graphic metaphor and to take verisimilitude as an analogue of falsifiability because Popper's formulation of the former notion is done also in logical terms.

43. We can see that this analogy holds in all important respects by noticing that the maximum degree of falsifiability is attributed to contradictory theories because they can be falsified by any possible state of affairs. But contradictory theories are not precisely an example of the good empirical theory that we are looking for. One can put this in less formal words by saying that a theory is better the higher its degree of falsifiability. So we want theories which by being easier to falsify have high empirical content, but we certainly do not want to formulate theories that are contradictory. There is a sense in which we want to exclude the extreme point of the continuum but keep it as a comparison point.

44. It should be patent why (12) furnishes a better topological representation of the degree of Vs of diverse (but related) theories than (6) and (7) above. On the other hand, Popper does not speak of 'absolutely false theories', and he reserves point zero for tautologies since their truth-content and falsity-content subclasses are empty; hence, they have no Vs. Orthodox Popperians may find the idea of absolute falsehood hard to swallow (after all, are not *all* falsehoods at *some* distance from the truth?) and they might object also to the removal of tautologies and self-contradictions from (12) or point out that (absolutely) false theories have a non-empty truth-content subclass. Granting that there is a tension between the basic implication of Vs (every false statement has *some* truth content) and my recommended treatment of absolutely false theories, I think my position is defensible on the following grounds: (i) it seems preferable to say that a theory comprised exclusively of false statements is at the maximal distance from the truth, as I suggest; (ii) my proposal is restricted to families of theories, in such a way that when we assign Vs zero to a theory t_x we only mean that t_x is at the maximal distance from the truth within that family (naturally, (11) and the conventions adopted for its sake, would not work to rank order unrelated theories, let alone the whole set of theories in empirical science).

45. This is the right time to point out an ambiguity that infects Popper's notion of verisimilitude. He usually defines this notion by appealing to the Tarskian operation of *logical consequence*, yet when he unpacks the definition and offers examples, he refers frequently to the *entailment relation*. But they are different. The former holds between sentences, whereas the latter is usually treated as a relation between propositions. More clearly: B is a logical consequence of A if in every model in which A is true B is true, or, equivalently, on every reinterpretation of the non-logical terms in A on which it is true, B is true as well. Accordingly, if B is a logical consequence of A, then A entails B, but the reverse need not be the case (i.e. it does not follow that if A entails B, then B is a logical consequence of A). One proposition entails another just in case the truth of the first is sufficient for the truth of the second, in the sense that its content and structure together with its truth guarantee the truth of the second given its content and structure. For example, let A be the proposition 'X is odd' and B 'X is not divisible by 2', then A entails B, but B is not a logical consequence of A. Kirk Ludwig (to whom I owe this point) has suggested to me that Popper probably needs to use entailment rather than logical consequence (and judging by some of Popper's examples – as well as my own examples below – this seems to be the case) because, in general, the content of a proposition should be identified with what it entails. To solve the problem that propositions which differ in content may have the same entailments (e.g. the proposition that C is a square and that C is a six-sided regular polyhedron) Ludwig recommends introducing a narrower notion of 'truth content', based on the entailment relation (e), which yields a more useful notion than would a similar definition in terms of logical consequence. Such definition can be stated in the following terms:

> S and R have the same (e)-truth content just in case everything that the proposition expressed by S entails the proposition expressed by R entails and vice versa.

46. It can be argued that the truth contents of these theories show nothing about their relative merits, because – as is known – several sets of distinct statements might agree with the same set of facts (or that evidence can corroborate even false theories). But this objection does not hit upon the logical characterization of Vs and is irrelevant to our present purposes.

47. In *RAS* Popper seems to give up any hope of finding a suitable measure of Vs, but contends that, for the purposes of his epistemology, it suffices with the intuitive use of the notion that grounds rationally our belief that a theory t_2 is closer to the truth than another theory t_1. The point is that

> though we may reasonably believe that the Copernican model as revised by Newton is nearer to the truth than Ptolemy's, there is no means of saying *how* near it is: even if we could *define* a metric for verisimilitude (which we can do only in cases which seem to be of little interest) we should be unable to *apply* it unless we knew the truth – which we don't. We may think that our present ideas about the solar system are near to the truth, and so they may be; but we cannot know it. (*RAS*, p. 61; see also pp. xxv–vii and 57–9)

48. Recall, however, that Popper does not want to treat *Vs* as a function of the mere cardinality of true and false statements but as a function of the truth-content and falsity-content subclasses.
49. Increase in truth content alone is not sufficient to guarantee an increase in *Vs*, and we need to consider also falsity content. Thus 'the only field left for scientific debate – and especially to empirical tests – is whether or not the falsity content has also increased. Thus our competitive search for verisimilitude turns, especially from the empirical point of view, into a competitive comparison of falsity contents (a fact that some people regard as a paradox)' (*OK*, p. 81).
50. While these objections appear to be devastating, I believe they can be satisfactorily met. Given the difficulties of arriving at a working metric for *Vs*, Popper has suggested that we should settle for an intuitive approach. On the other hand, he also recommends us to stay away from unnecessary complexities and technicalities. While this may discourage friends of precision, one needs to keep in mind that judgements of verisimilitude are not demonstrable (not even with the help of a suitable metric, were one developed)

> but may be nevertheless asserted as a conjecture, strongly arguable for or against on the basis of (1) a comparison of the logical strength of the two theories and (2) a comparison of the state of their critical discussion, including the severity of tests which they have passed or failed. ((2) can also be described as a comparison of their degree of corroboration.) (Popper 1976a p. 158; see also *RAS*, p. 277)

For a defence of the intuitive cogency of the idea of verisimilitude see Koertge 1978.

51. I have, deliberately, omitted all reference to the continued work of many authors (after Tichý and Miller) on the theory of verisimilitude. Some of these authors have considered the objections to Popper's qualitative theory of verisimilitude discussed here, but I have not examined their replies because, as announced before, I have preferred an intuitive approach to the problem. The interested reader should see: Brink, C. 1989, 'Verisimilitude: views and reviews', *History and Philosophy of Logic* 10, pp. 181–201; Britz, K. and Brink, C., 1995, 'Computing verisimilitude', *Notre Dame Journal of Formal Logic* 36, pp. 30–43; Niiniluoto, I., 1985, 'The significance of verisimilitude', *Philosophy of Science Association* 2; Niiniluoto, I., 1987, *Truthlikeness*, Dordrecht: D. Reidel; Niiniluoto, I., 1998, 'Verisimilitude: the third period', *British Journal for the Philosophy of Science*, 49, pp. 1–29; Oddie, G., 1986, *Likeness to Truth*, Dordrecht: D. Reidel.

Appendix
List of Definitions

Popper has often modified his definition of key notions. In this list, I have included the version that I think best supports a charitable reading of Popper's account of science.

Basic statement (also **test-statement**): a statement that can serve as premise in an empirical falsification.

Corroboration (**degree of**): the degree to which a hypothesis has stood up to tests.

Event: a set of occurrences of the same kind.

Empirical content (also **informative content**): The amount of empirical information conveyed by a statement or a theory. Its degree is determined by the 'size' of the class of potential falsifiers.

Falsifiability (or **testability**): the logical relation between a theory and its class of potential falsifiers. Falsifiability is a criterion of the empirical character of a system of statements.

Falsification: the conclusive demonstration that a theory has clashed with a falsifier.

Falsity content (of x): the subclass of false consequences of x. It is not a Tarskian consequence class.

Occurrence: a fact described by a singular (basic) statement.

Logical content (of x): the class of (non-tautological) statements entailed by x, where x can be a statement or a theory. A Tarskian consequence class.

Logical probability (of x): the probability of x relative to some evidence; that is to say, relative to a singular statement or to a finite conjunction of singular statements.

Logical strength: increases with content (or with increasing *improbability*).

Truth content (of x): the class of (non-tautological) true logical consequences of x. It is a subclass of the logical content.

Verisimilitude (of x): the degree of closeness to the truth of x. It can be measured as the difference of truth content minus falsity content.

References

Ackermann, Robert, 1976, *The Philosophy of Karl Popper*. Amherst, MA: University of Massachusetts Press.
Agassi, Joseph, 1964, 'The nature of scientific problems and their roots in metaphysics', in Bunge (ed.) 1964.
Agassi, Joseph, 1968, 'The novelty of Popper's philosophy of science', *International Philosophical Quarterly* 8.
——, 1972, 'Imperfect knowledge', *Philosophy and Phenomenological Research* 32.4.
——, 1981, 'To save verisimilitude', *Mind* 90.
——, 1983, 'Criteria for plausible arguments', *Mind* 74.
——, 1993, *A Philosopher's Apprentice: In Karl Popper's Workshop*. Amsterdam: Rodopi.
——, 1994, 'An inductivist version of critical rationalism', *Philosophy of the Social Sciences* 24.4.
Agassi, Joseph (ed.), 1982, *Scientific Philosophy Today: Essays in Honor of Mario Bunge*. Dordrecht: Reidel.
Amsterdamski, Stefan, 1996, *The Significance of Popper's Thought*. Amsterdam: Rodopi.
Ayer, Alfred, 1956, *The Problem of Knowledge*. London: Macmillan.
——, 1974, 'Truth, verification and verisimilitude'. In Schilpp (ed.), 1974.
Barker, Stephen F., 1957, *Induction and Hypothesis: A Study of the Logic of Confirmation*. Ithaca, NY: Cornell University Press.
Bartley III, William, 1964, 'Rationality versus the theory of rationality', in Bunge (ed.) 1964.
Bunge, Mario (ed.), 1964, *The Critical Approach to Science and Philosophy*. New York: Free Press of Glencoe.
Burke, T.E., 1983, *The Philosophy of Popper*. Manchester: Manchester University Press.
Chalmers, A.F. 'On learning from our mistakes'. *British Journal for the Philosophy of Science* 24.
Chapin, Seymour L., 1957, 'The astronomical activities of Nicolas Claude Fabri de Pereisc', *Isis* 48.1.
Churchland, Paul M., 1975, 'Karl Popper's philosophy of science', *Canadian Journal of Philosophy* 5.
Cohen, Jonathan, 1977, *The Probable and the Provable*. Oxford: Clarendon Press.
——, 1980, 'What has science to do with truth?' *Synthese* 45.
Curry, Gregory and Musgrave, Alan (eds), 1985, *Popper and the Human Sciences*. Dordrecht: Martinus Nijhoff.
D'Amico, Robert, 1989, *Historicism and Knowledge*. New York: Routledge.
Derksen, A., 1985, 'The alleged unity of Popper's philosophy of science: falsifiability as fake cement', *Philosophical Studies* 48.

Drake, Stillman (ed.), 1957, *Discoveries and Opinions of Galileo*. Garden City, NY: Anchor Books.

Duhem, Pierre, 1954, *The Aim and Structure of Physical Theory*. Princeton, NJ: Princeton University Press.

Fain, Haskell, 1961, Review of *The Logic of Scientific Discovery*, *Philosophy of Science* 28.3.

Feigl, Herbert, 1981, *Inquiries and Provocations: Selected Writings 1929–1974*. Dordrecht: Reidel.

Feyerabend, Paul, 1968, 'A note on two "problems" of induction', *British Journal for the Philosophy of Science* 19.

——, 1975, *Against Method*. London: New Left Books.

Galileo, 1961, *The Sidereal Messenger*. London: Dawsons.

García, Carlos E., 1998, *Historical Evolution of Scientific Thought*. Manizales: University of Manizales.

Groth, Miles, 1997, 'Some precursors of Popper's theory of knowledge', *Philosophy in Science* 7.

Grünbaum, Adolf, 1976, 'Is the method of bold conjectures and attempted refutations justifiably the method of science?', *British Journal for the Philosophy of Science* 27.

Gupta, Chhanda, 1993, 'Putnam's resolution of the Popper–Kuhn controversy', *Philosophical Quarterly* 43.172.

Hacohen, Malachi H., 2000, *Karl Popper: The Formative Years 1902–1945*. New York: Cambridge University Press.

Harris, Errol, 1972, 'Epicyclic Popperism', *British Journal for the Philosophy of Science* 23.

Harris, John, 1974, 'Popper's definitions of "verisimilitude".' *British Journal for the Philosophy of Science* 25.

Hempel, Carl, 1992, 'Empiricist criteria of cognitive significance: problems and changes', in Boyd and Trout (eds), *The Philosophy of Science*. Cambridge, MA: MIT Press.

Howson, Colin, 1984, 'Popper's solution to the problem of induction', *Philosophical Quarterly* 34.135.

Hume, David, 1978, *A Treatise of Human Nature*. Oxford: Oxford University Press.

——, 1993, *An Enquiry Concerning Human Understanding*. Indianapolis, IN: Hackett.

Johansson, Ingvar, 1975, *A Critique of Karl Popper's Methodology*. Stockholm: Scandinavian University Books.

Jones, K.E., 1973, 'Verisimilitude versus probable verisimilitude', *British Journal for the Philosophy of Science* 24.

Kampits, Peter, 1980, 'The myth of reason and critical rationalism', *Listening* 15.

Koertge, Noretta, 1978, 'Towards a new theory of scientific inquiry', in G. Radnitzky and G. Andersson (eds), *Progress and Rationality in Science*. Dordrecht: Reidel.

Kuhn, Thomas, 1962, *The Structure of Scientific Revolutions*. Chicago, IL: University of Chicago Press.

——, 1974, 'Logic of discovery or psychology of research', in Schilpp 1974.

Lakatos, Imre, 1970, 'Falsification and the methodology of scientific research programmes', in Lakatos and Musgrave (eds), 1970.

——, 1974, 'Popper on demarcation and induction', in Schilpp 1974.

——, 1981, 'History of science and its rational reconstructions', in Ian Hacking (ed.), *Scientific Revolutions*. Oxford: Oxford University Press.

Lakatos, Imre and Musgrave, Alan (eds), 1970, *Criticism and the Growth of Scientific Knowledge*. Cambridge: Cambridge University Press.

Laudan, Larry, 1981, 'A confutation of convergent realism', *Philosophy of Science* 48.1.

Levison, Arnold, 1974, 'Popper, Hume, and the traditional problem of induction', in Schilpp 1974.

Levinson, Paul (ed.), 1982, *In Pursuit of Truth: Essays on the Philosophy of Karl Popper on the Occasion of his 80th Birthday*. New York: Humanities Press.

Lewis, David, 1985, 'Languages and language', in A.P. Martinich (ed.), *The Philosophy of Language*. New York: Oxford University Press.

Lindberg. D.C., 1983, *Studies in the History of Medieval Optics*. London: Variorum Reprints.

Magee, Bryan, 1973, *Karl Popper*. New York: Viking Press.

Maxwell, Grover, 1974, 'Corroboration without demarcation', in Schilpp 1974.

McEvoy, John, 1975, 'A "revolutionary" philosophy of science: Feyerabend and the degeneration of critical rationalism into sceptical fallibilism', *Philosophy of Science* 42.

Miller, David, 1974, 'Popper's qualitative theory of verisimilitude', *British Journal for the Philosophy of Science* 25.

——, 1976, 'Verisimilitude redeflated', *British Journal for the Philosophy of Science* 27.

——, 1999, 'Popper and Tarski', in Ian Jarvie and Sandra Pralong (eds), *Popper's Open Society: After Fifty Years*. New York: Routledge.

——, 1994, *Critical Rationalism: A Restatement and Defense*. Peru: Open Court.

Motterlini, Matteo (ed.), 1999, *For and against Method. Imre Lakatos and Paul Feyerabend*. Chicago, IL: University of Chicago Press.

Musgrave, Alan, 1975, 'Popper and "diminishing returns" from repeated tests', *Australasian Journal of Philosophy* 53.

Narayan, S. Shankar, 1990, 'Karl Popper's philosophy of scientific knowledge', *Indian Philosophical Quarterly* 1.

Niiniluoto, Ilkka, 1982a, 'What shall we do with verisimilitude?', *Philosophy of Science* 49.

——, 1982b, 'Truthlikeness for quantitative statements', *Philosophy of Science Association* 1.

——, 1985, 'The significance of verisimilitude', *Philosophy of Science Association* 2.

Notturno, Mark, 1985, *Objectivity, Rationality, and the Third Realm: Justification and the Grounds of Psychologism*. Boston, MA: Kluwer.

Oddie, Graham, 1990, 'Verisimilitude by power relations', *British Journal for the Philosophy of Science* 41.

O'Hear, 1975, 'Rationality of action and theory-testing in Popper', *Mind* (new series), 84.334.

——, 1980, *Karl Popper*. London: Routledge & Kegan Paul.

O'Hear, Anthony (ed.), 1995, *Karl Popper: Philosophy and Problems*. New York: Cambridge University Press.

Popper, Karl, 1950, *The Open Society and its Enemies*. Princeton, NJ: Princeton University Press.

——, 1955, 'A note on Tarski's definition of truth', *Mind* 64.255.
——, 1957, *The Poverty of Historicism*. Boston, MA: Beacon Press.
——, 1959, *The Logic of Scientific Discovery*. London: Hutchinson.
——, 1963, *Conjectures and Refutations: The Growth of Scientific Knowledge*. London: Routledge & Kegan Paul.
——, 1968, 'Remarks on the problems of demarcation and of rationality', in Imre Lakatos and Alan Musgrave (eds) *Problems in the Philosophy of Science*. Amsterdam: North Holland Publishing.
——, 1972, *Objective Knowledge: An Evolutionary Approach*. Oxford: Clarendon Press.
——, 1974, 'Replies to my critics' in Schilpp 1974.
——, 1975, 'The rationality of scientific revolutions', in Rom Harré (ed.), *Problems of Scientific Revolution*. Oxford: Oxford University Press.
——, 1976a, 'A note on verisimilitude', *British Journal for the Philosophy of Science* 27.
——, 1976b, *Unended Quest: An Intellectual Autobiography*. La Salle: Open Court.
——, 1982a, *Quantum Theory and the Schism in Physics*. London: Hutchinson.
——, 1982b, *The Open Universe: An Argument for Indeterminism*. London: Hutchinson.
——, 1983, *Realism and the Aim of Science*. London: Hutchinson.
——, 1990, *A World of Propensities*. Bristol: Thoemmes.
——, 1992, *In Search of a Better World: Lectures and Essays from Thirty Years*. New York: Routledge.
——, 1994a, *Knowledge and the Body–Mind Problem: In Defence of Interaction*. New York: Routledge.
——, 1994b, *The Myth of the Framework: In Defence of Science and Rationality*. New York: Routledge.
——, 1998, *The World of Parmenides: Essays on the Presocratic Enlightenment*. New York: Routledge.
——, 1999, *All Life is Problem Solving*. New York: Routledge.
Popper, Karl and Miller, David, 1983, 'A proof of the impossibility of inductive probability', *Nature* 302.
——, ——, 1987, 'Why probabilistic support is not inductive', *Philosophical Transactions of the Royal Society of London* (Series A, Mathematical and Physical Sciences), 321.1562.
Post, John F., 1972, 'The possible liar', *Noûs* 4.4.
——, 1972, 'Paradox in critical rationalism and related theories', *Philosophical Forum* 3.
Putnam, Hilary, 1974, 'The "corroboration" of theories', in Schilpp (ed.) 1974.
Radnitzky, Gerard (ed.), 1987, *Evolutionary Epistemology*. La Salle: Open Court.
Russell, Bertrand, 1990, *The Problems of Philosophy*. Indianapolis, IN: Hackett (repr. of the 1912 edn).
Schilpp, Arthur (ed.), 1974, *The Philosophy of Karl Popper*. La Salle, IL: Open Court.
Simkin, Colin, 1993, *Popper's Views on Natural and Social Science*. Leiden: E.J. Brill.
Stenect, N. and Lindberg, D.C., 1983, 'The sense of vision and origins of modern science', in Lindberg 1983.
Stokes, Geoffrey, 1998, *Popper: Philosophy, Politics and Scientific Method*. Malden: Polity Press.
Stove, David, 1982, *Popper and After*. Oxford: Pergamon Press.

Tichý, Pavel, 1974, 'On Popper's definitions of verisimilitude', *British Journal for the Philosophy of Science* 25.
——, 1976, 'Verisimilitude redefined', *British Journal for the Philosophy of Science* 27.
Warnock, G.J., 1960, Review of *The Logic of Scientific Discovery*, *Mind* 69.273.
Watkins, John, 1984, *Science and Scepticism*. Princeton, NJ: Princeton University Press.
——, 1995, 'How I almost solved the problem of induction', *Philosophy* 70.
Wettersten, John R., 1992, *The Roots of Critical Rationalism*. Amsterdam: Rodopi.
——, 1999, 'New problems after Popper', *Philosophy of the Social Sciences* 29.1.
Wittengenstein, Ludwig, 1981, *Tractatus Lógico-Philosophicus*. London: Routledge & Kegan Paul.

Index

absolute empirical truth   22, 24, 128
absolute falsehood   143, 153n. 44
absolute logical probability   115n. 15, 148n. 13
absolutely false theories   144, 153n. 44
absolutely true theories   143, 144
acceptance   15, 17, 19, 54, 63, 90
Ackermann, R.   77, 114n. 7
  on falsifiability   77n. 24.
Agassi, J.   75n. 11, 150n. 22
  on falsifiability   75n. 10.
aim of science   6, 24, 91, 107, 121, 140, 149n.18
Amsterdamski, S.   157
approximation to the truth   19, 73, 121, 124, 130 *see also* truthlikeness and verisimilitude
apriorism   13–24
Aristotle   61, 81n. 58
auxiliary hypotheses   41, 60, 69
Ayer, A.J.   29, 30, 37n. 39–40, 75n. 10, 119n. 40
  on Popper   29

Bacon, F.   65
background knowledge   31, 59, 60, 67, 94–5, 130
Barker, S.   118n. 38
Bartley, W.W., III   116n. 19
basic statement   44–8, 53, 59, 63–4, 85, 88–9, 94, 103, 105, 109, 139
Burke, T.E.   7n. 2, 37n. 40

Carnap, R.   114n. 5
Chapin, S.L.   82n. 76
closeness to the truth   124, 129 *see also* verisimilitude

Copernicus   66–8, 146
Cohen, J.   32, 34, 114n. 4
  on Popper   32–4
comprehensively critical rationalism   116n. 19
comprehensively true theory   124, 128, 129, 153n. 41
confirmation   64, 91, 103, 113n. 1
conjecture and refutation   7, 20, 124
consequence class   124–5, 127, 148n. 8, 149n. 17, 152n. 38
context of discovery   13–14
context of justification   13–14
context of practice   32
conventionalism   57, 63, 122
correspondence theory of truth   122
corroborability   4, 89, 96–8, 101, 111
  and empirical content   118n. 31
  and high falsifiability   112, 113n. 2, 118n. 34
  and logical probability   116n. 20, 127
  and probability   117n. 25, 127
  and testability   118n. 31
corroborated   68, 70–3, 84, 88–91, 98–101, 121
corroboration   5, 84–120, 130–1, 146–7
  as an appraisal   88, 90, 109
  and basic statements   90
  and content   100
  criticisms of   101–13
  degree of   21, 30, 88–90, 93–4
  and failed falsification   111
  and falsifiability   115n. 10, 16
  how to determine the degree of   88, 90
  of a hypothesis   94, 95, 102
  and induction   102, 105, 106
  and logical probability   89, 91, 93

Index

measures of 95
negative 117n. 30
negative degree of 89, 96
notion of 84, 88, 92–101
positive degree of 88, 90, 96
and probability of events 85
the problem of 92
and the problem of induction 87, 92, 101, 108
and repeated tests 101
of self-contradictions 89, 96, 98, 99
and test statements 88
and truth-value 90
and verification 102
and verisimilitude 107, 109, 110, 111
of tautologies 89, 95, 96, 98, 99
of a theory 87, 98, 99, 106
corroborating instances 88–90, 115n. 15
critical discussion 33, 41, 65, 70, 111
criticizability 116n. 19
critical rationalism 1–2, 66, 68

D'Amico, R. 82n. 71
  on fallibilism 79n. 48
deductive logic 39, 44
deductivism 5, 38, 109
  and demarcation 44
  and the logic of science 39
degrees of universality 129
demarcation 5, 9, 15, 18, 38, 40–3, 51–3, 57–9, 64, 107, 109
  criterion of 9, 57, 59
  and meaning 9, 41–2, 44
  problem of 18
Derksen, A. 5, 111, 112, 120n. 59, 131
  on Popper 70–4, 82n. 78–9
disjunctive normal form 133–4, 142, 145
distance to the truth 132, 148n. 9, 148n. 12 *see also* verisimilitude
Drake, S. 81n. 58
Duhem, P. 60, 80n. 56

Einstein, A. 88, 117n. 24
empirical basis 59, 63, 139
empirical content 124, 128, 139, 140

empirical import 128
empirical science 14–15, 17, 30, 32, 38–41, 44–5, 49–50, 106, 121, 124–6, 143–4
empirical success 130, 150n. 22
empirical testing 130
empirical theories 4, 9, 13, 38, 50, 99, 128
entailment relation 154n. 45
epistemic value 145

Fain, H. 34n. 1
fallibilism 55, 58–9, 63
false statement 126, 129, 142
false theory 129, 130, 133, 134, 135, 137, 143
falsifiability 4, 9, 38–83, 88, 98, 100
  and auxiliary hypotheses 41, 60
  and deduction 38–45
  definition of 44–50
  degrees of 38, 89, 96, 97
  and demarcation 40, 42
  distinct notions 71–3
  and empirical content 48
  and falsification 45–6, 56
  and laws of nature 42
  maximum degree of 153n. 43
  and verifiability 41–3
  and verification 41–3
falsification 9, 15, 18, 28, 30–1, 38–9, 41–6, 49–51, 53, 55–9, 64–7, 70, 73, 89, 97
  and truth 131
falsified theories 97, 107, 118n. 33, 129, 130
falsity content 124, 126, 127, 130, 131, 134, 135, 136, 137, 139, 155n. 49
  measure of 124, 133, 141
falsity-frequency 86
families of theories 152n. 41, 153n. 44
Feigl, H. 37n. 47
Feyerabend, P. 2, 5, 37n. 44, 51, 64–70
  and the history of science 82n. 73
  and irrationalism 64
  on Popper 51, 64–70

Galileo, G. 58, 151n. 28
  astronomical discoveries 58, 60–2, 66–70, 81n. 58–9, 82n. 76
García, C.E. 82n. 73
Goodman, N. 31
  paradox 31, 37n. 44
Gupta, Ch. 79n. 47
  on Popper and Kuhn 79n. 47

Harris, J. 150n. 22
Hempel, C. 41
  on falsifiability 41, 75n. 10
Howson, C. 24–7
  on Popper 9, 24–8, 36–7
Hume, D. 4, 8, 10–12, 22–3, 32, 102
  analysis of causation 11, 35n. 4_{ff}
  on induction 8, 9, 10–12, 35n. 4_{ff}

induction 4, 7
  the problem of 4, 8, 10, 15–23, 26
  and technology 32–3
inductive logic 8, 13, 20
  and probability 87, 92, 116n. 23
inductivism 9, 26,
initial conditions 94, 105
infinite regress arguments 7, 40, 63, 87
informative content 113, 125, 128, 134, 139, 140
instrumentalism 121
irrationalism 7, 16–17, 19, 24, 64, 71, 108
isolated existential statements 41, 125

Johansson, I. 7n. 2, 76n. 21
Jones, K.E. 158

Kepler, J. 68–9, 103, 106
Koertge, N. 155n. 50
Kuhn, T.S. 2, 104, 119n. 44
  on falsifiability 51–7, 79n. 39
  on the logic of discovery 53–4
  on realism 78n. 37
Lakatos, I. 2, 107, 109, 110, 131, 150n. 22
  on falsifiability 51, 57–64, 79n. 49, 80n. 50, 55, 57
  on methodological falsificationism 108

language dependence 141, 151n. 31, 152n. 37
law of diminishing returns 6, 30
law of nature 21–3, 28, 33
learning from our mistakes 20, 123
Levison, A. 29
  on Popper 29, 36n. 30, 37n. 44
Levinson, P. 37n. 34
Lindberg, D.C. 81n. 59
logical consequence 95, 124, 135, 154n. 45
logical content 124, 126, 134
  measure of 125
logical improbability 94, 103, 119n. 45, 141
logical operations 137, 138, 139
logical probability 133, 134
logical strength 133, 134, 141
Ludwig, K. 154n. 45

Magee, B. 76n. 17
Maxwell, G.
  on Popper 36n. 30
McEvoy, J. 159
metaphysical 18, 22–3, 28–9, 40, 42–3, 48–9, 59
metaphysics 9, 108
Miller, D. 28, 32, 116n. 23, 121, 132, 134, 135, 136, 137, 139, 141, 147n. 5, 151n. 28, 31
  on Popper 28, 37n. 34
Motterlini, M. 80n. 55, 81n. 64
Musgrave, A. 37n. 43

Narayan, S. 76n. 21
Newton, I. 22, 23, 117n. 24, 151n. 28, 154n. 47
Niiniluoto, I. 155n. 51
Notturno, M. 159
objective knowledge 16, 21
objective truth 121, 122, 123

Index

objectivity 38, 119n. 50
Oddie, G. 159
O'Hear, A. 2, 28, 30, 31, 118n. 37
  criticism of Popper 28–9, 30–1

Popper, K.R. 1
  against justificationism 9, 19, 20
  against psychologism 13–14, 21
  on the asymmetry of verification and falsification 5, 17, 41–3
  on the content of science 48–9, 74, 89–91, 93–4, 97
  on corroboration 84–101
  on critical rationalism 19, 20, 66–8
  critics to the positivists 8
  demarcation criterion of 9
  on election between theories 33
  on existential statements 41–3, 75n. 12
  on falsifiability 44–50, 74n. 1, 75n. 11, 76n. 16–17, 77n. 22, 78n. 28
  on falsification 19, 30, 45–6
  formulation of the problem of induction 8–10, 15–16, 18, 21
  frequency interpretation of probability 85–7
  on Goodman's paradox 31–2
  and Hume on induction 16–17, 18, 20, 36n. 30
  and irrationalism 19
  and Kuhn 52, 78n. 36, 79n. 42
  on laws of nature 21–3
  on metaphysics 9, 22, 23, 28
  on predictions and prophecies 9, 10, 34
  on probability of hypothesis 24, 32
  on the rationality principle 111
  on scientific progress 6, 9, 17
  on severity of tests 30, 31, 88–92
  replies to Kuhn 54–5
  solution to the problem of induction 4, 9, 12–34, 36n. 29
  on the success of science 107, 123
  theory of science 1–2
  theory of verisimilitude 124–31
  on truth 17, 19–20, 23, 55, 84, 121–3

positivism 1, 8–9, 40–1, 43
Post, J. 116n. 19
potential falsifier 44–50, 139
precision 129
preferences among theories 33–4, 64, 73, 84–5, 106, 112
principle of empiricism 15, 17, 20, 40
principle of verifiability 9, 17, 41–3
  see also verification
probabilistic support 116n. 23
probability 9, 12, 13, 24–6, 32, 85–92
  of events 85–7
  frequentist interpretation of 85
  of hypotheses 85–7
  of occurrences 85
  of a statement 85
progress in science 9, 17, 20–1, 39, 52, 59–60, 64–6, 71, 73, 91, 107–9, 122
Ptolemy 68, 145
Putnam, H. 2, 102, 103, 104, 105, 107, 119n. 39, 119n. 41, 44, 45

Radnitzky, G. 116n. 19
rationalism 1–2, 9, 66, 68
rationality principle 60
reasons 16, 18–20, 25–6, 31, 33–4, 63
Reichenbach, H. 85, 114n. 6, 7
repeatability of tests 30
reproducible effects 31, 46–7
Russell, B. 18

scepticism 35n. 12, 79n. 48
  about science 19, 26
Schilpp, A. 107, 76n. 15
science 1, 8–10, 15–17, 24, 30, 34
science and pseudoscience 24, 109, 121
scientific realism 1, 78n. 37
scientific statements 38, 59, 91, 103
  as tentative conjectures 91, 102
Simkin, C. 119n. 50
simplicity 71, 88, 130
Stenect, N. 160
Stokes, G. 118n. 37
Stove, D. 7n. 2
subjectivist epistemology 123
support 8, 12, 32, 40, 61, 64, 68

tautological operations 76n. 20, 138
Tarski, A. 108, 110, 122, 135, 141,
     147n. 4, 148n. 6
tarskian consequence class 124, 148n. 8,
     149n. 17, 152n. 38
technology 32–3
testability 43–4, 52, 71–2, 74, 87–9, 91,
     94, 101, 130, 150n. 21
   degree of 71, 88, 89, 94, 96
tests 10, 14, 16–17, 19, 20–1, 29–31, 33
   severity of 30, 39, 88, 117n. 26,
     118n. 35
test situation 137, 138
test statement 139, 150n. 22, 23, 24,
     151n. 25, 27, 67
Tich, P. 132, 133, 134, 136, 137, 139,
     141, 152n. 33
trivialities 36n. 33, 75–6n. 12
true 12, 13, 16–20, 24–5, 33, 45, 87, 90,
     121–2
   by convention 59–60, 62
   statement 126, 129, 142
   theory 129, 135, 143, 149 n. 18
truth 6, 14–16, 19, 25–6, 28, 55, 71, 84
   and content 124
   excess of 137, 138
   maximal 126
   measure of 124
truth-content 124, 126, 127, 130, 131,
     134, 135, 136, 137, 138, 139
truth-frequency 85–6
truth-value 41, 67, 81n. 61, 90, 148n. 13
truthlikeness 17, 108, 124, 130, 137, 144
   see also verisimilitude

uniformity of nature 28–9

verification 5, 9, 29–30, 41–3, 85, 100,
     103
verificationist 5, 14, 55, 101

verisimilar 121, 141, 143
verisimilitude 4, 6, 17, 26, 108, 112,
     118n. 32, 120n. 58, 121–55
   appraisals of 142
   as applied to false theories 125
   of an arbitrary sentence 134
   and background knowledge 130
   as a comparative notion 143
   comparison by 136
   the conjectural status of 130, 131
   as a conjecture 155n. 50
   and contradictions 128
   and corroboration 131
   definition of 124–31
   degrees of 122, 123, 124, 128, 143,
     144
   and induction 131
   logical definition of 132, 149n. 14
   maximum 148n. 11
   measure of 131
   the notion of 124
   probabilistic definition of 133
   problems of the qualitative definition of
     131, 134, 138
   problems of the quantitative definition
     of 131
   the quantitative definition of 141
   solution of the qualitative definition of
     138, 139
   and tautologies 126, 127, 128
   of trivial statements 125, 140

Warnock, G.J. 118n. 37
   on Popper 34n. 2
Watkins, J. 118n. 37
   on Popper 35n. 3
Wettersten, J. 161
Wittgenstein, L. 148n. 13